Bill and Marcia,
I do hope you will enjoy this.
In part, it rose out of my preparation
for the Am. Farm Project.

Love,
Joe

Handful of Thunder:

A Prairie Cycle

Joe Paddock

ISBN 0-918552-15-X

Anvil Press
Box 37, Millville, MN 55957

This book is dedicated to the well-being of the prairies
of the northern United States and to
those beings who inhabit them.

ACKNOWLEDGEMENTS

Within this poem there are a number of short quotes from oral history which have been printed in italics. Most of these selections came from interviews I did myself and were originally printed in a book I edited entitled *The Things We Know Best,* an oral history of Olivia, Minnesota, and its surrounding countryside. In the main, this material has been given exactly as spoken by the interviewees, but I have here and there made some adjustments. My thanks to Book 200, Inc. of Olivia for their permission to reprint these selections. I also want to thank those "students" and interviewees who once worked with me on a project which resulted in the booklet *Milroy Memories* for their work and a brief quote which I drew from that publication.

Joe Paddock

INTRODUCTION

We yearn to be reunited, reconciled with the land. For our civilization has not merely abused this mother of our existence; it has done something more serious: it has ignored her, treated her as if she were irrelevant. We know, while hiding this guilt even from ourselves, that there will be retribution. We will pay for our sins, as will our children, to the fourth generation and beyond. Yet finally, we will have to come back. As the poet Meridel LeSueur has said, we can beat on our mother's breast in rebellious rage, but we still have to suck.

In this extraordinary epic, Joe Paddock has portrayed our assault upon the prairie, where meadowlarks now sing only in ditches. He has forced us to see and feel the violated flesh of the victim, her body laid open to the tearing of wind and rain.

Stung by these images, we cry for the land, and for ourselves.

With a vision as expansive as the prairie itself, Joe Paddock tells the story of the heartland: the receding glaciers. . .the migration of buffalo and Indians onto the new grassland. . .the white invasion, slaughtering buffalo and Indian. . .the white tide of settlement. . .the conquest of the white culture by chemicals and machines. A cycle of death.

When he writes of the prairie, Paddock is speaking not only of this great, rich expanse in the center of our country. He is speaking of all the land, of all the earth, of creation itself. Yet he tells the prairie's story with a particular and detailed knowledge earned through years of living intimately with it. Growing up at the edge of the prairie in Litchfield, Minnesota, (where with his wife, the poet Nancy

Paddock, he has recently returned to live) he knew both the gentleness and the wrath of the prairie; he knew the terror one can feel in her presence. During his early manhood in the 1960s, he made his living by gathering specimens and trapping, "reaching into Her lap for a round of sweet blood-soaked fruit." He has experienced the blood lust and the guilt from that taking.

How then do we come back to our mother, reconciled? By becoming real: by embracing the pain of our knowledge and turning our faces to the light, embodied in "Christ-God-Buddha, wherever enlightenment lives. . .that we might begin/ to correct our days,/ and pray/ we are not too late. . .to sort out, to seek and find/ the holy balance."

Joe Paddock speaks in a voice that somehow manages to avoid the alienation of academic literature and yet raise common experience to the power of art. He does this out of a firm rootedness in the land and culture of Minnesota and the Midwest and through hard years of literary apprenticeship. With this haunting and inspiring work, I believe he will emerge as an important new folk poet in the tradition of Walt Whitman and Meridel LeSueur.

Jack Miller
Editor, *North Country Anvil*

Table of Contents

I. The White Bubble

1

This vastness of grassland gone
to corn and beans is a tiny thing
under the nighttime leaping prairie fire
of galaxies. Yes, the entire universe
is involved, everything
beaming good influence through
everything else. The mind receiving
becomes calm
enough to begin:

for ten thousand years
these prairies had been living
balance, hovering, sensitive as
a white bubble over black water
under ice. A balance,
the grasses rising even
as that last glacier sank back
to its pole, and the buffalo
followed the grass, and wolves
and red men and women, living deep

in the balance, followed the buffalo,
and the great flights
of swans and geese and cranes
followed the seasons. . . .
Finer and finer: sunlight, trace element,
microbe, mist, nematode, mole. . . . O,
the eyes, even of red men, cannot
follow the claw marks of electrons,
atoms, ions, starlight, all
radiations and migrations infinitely seeking,
reaching   balance.

2

Ten thousand years,   and then
the white bubble slipped, leapt
almost clear of its pond.
A new force, the hard half
of the human race,
pressed, worn, drawn inexorably toward
another great expanse of pure
nature.   How
could they resist? The polarity
was too great, and they swept west,
swept ten thousand years of prairie
clean
down to its grass
in about one life
the antelope and the buffalo, the red men,
women and children   were sent back down
into the seething

ground   or held
in reserves.

3

Buffalo hunters stank, they say,
so that, downwind, your gorge grew
to a filthy fist   in your throat.
The stink of carcasses, those years
must have been   buzzard heaven.
Buzzards must have fallen
drunk from the skies.

That was the last,
that great maggot squirm, that glut
(we're talking, now, only a bit
more than one long lifetime back)
was the end, the buzzard feast
which began—

"Well, how in hell you gonna string a fence
with sixty million buffalo"
thundering on the balance?
"How in hell
you gonna run a railroad
with sixty million buffalo"
thundering on a balance
which has so little to do
with the market?

Land has to be cross-hatched
and plotted   for white men
to sleep easy on it.
White men can't sleep on land uncut
by fences, fields, roads,
power lines, phone lines, gas lines, water mains. . . .
They are strung   from his part
of the brain.   They trick him
into a sort of ease,
are the mesh on which his mattress swings.

No room, then, for the immense
migrations of matted and hooved tons.
Buffalo guns, heavy-barreled Sharps—
on rest sticks, from train windows—
boomed a somber dirge, and each
single buffalo coughed bright blood
into the light, sank to its knees
and down   into grass. The great
herds receded like glaciers. Great
innocent carcasses sank back in
to roots   like melting ice.

4

The prairie lay clean
and waiting.   O,
there were still the bones,
the bones yet to be picked.
That wondrous empty galaxy of grass
turned, for twenty years, one vast

boneyard.
White men, bonepickers, combed
buffalo grass and bastard grass, bent
nearly double, as if they had lost
in childhood, some favorite toy,
an animal, perhaps, they had hugged
in their sleep.

Hounding bones. Wagonloads of bones,
four dollars to the ton,
to be used in China
teacups   and
to fertilize the ground.
                                        (As if they wouldn't have
                                        on their own.)

5

The prairie lay clean,
and the great white way streamed west,
instantaneous (in galaxy time)
as a chemical reaction:
they came like charged hydrogen
after oxygen   that it might rest
as water. Seeking the level
which will give rest, ease.
Nature's abhorrence
of what had been achieved—
that great emptiness — sucked in the guts
and minds of poor European peasants.
An emptiness which translated

into place and plenty. They swept west,
pulling roots which ran deep
into the Middle Ages. Roots
once set tenderly and with ritual
to the moon's phases, under the radiance
of Her full belly.

They came, flowing
out of the Middle Ages, the infinite
past, another of the great
migrations, like red men
thousands of years before them.
They came,
swept west like rumbling
herds, like great flights, like sunlight,
like galaxies streaming.
They came, pulled
by the emptiness, the hollow moaning
in their bellies
for the opposite pole.
Hope pulled them, hunger
and greed   and lies
in their minds,   on the tongues
of speculators, ads in great newspapers:

> "This is the sole
> remaining section of paradise
> in the western world; all the wild romances
> of the gorgeous Orient
> dwindle into nothing
> when compared to the everyday realities
> of Dakota's progress."

6

They came and found
swept emptiness.
The women saw it
lonely, an immense bowl
of madness, but the men went first
and saw only that gold-glinting haze
over grass—*fortunes in wheat!*

> *Well, we got out to the new farm*
> *that evening around sunset.*
> *We had come so far*
> *to be there, and the first thing*
> *my mother said was, "Why*
> *did you bring us to such*
> *a God-forsaken place as this?"*
> *"Well," my father says, "Well,*
> *you are my family.*
> *We're goin' to try to live together*
> *and make it here."*

II. The Body with a Hundred Hands

1

The women saw it, knew
that a dream
of a fortune in wheat
is not a life.

They'd been promised a return
to Eden
and wanted a garden
and a town, all a round
dance, human hand joined
to human hand. (Not so different from
the balance
of buffalo, wolves and grass.)
And it all, for a moment, happened,
rose (along the grid)
with the same yearning force
of first wheat: small
school, small church, small town. . . .
Nature feeding Her force through the splendid
forked creature, runners springing   out from roots

of the human plant. New kids,
fleshed from new ground,
barefoot with rootlets beginning
from tiny toes   down
into rich earth, repository and source
of it all   to that point,
the whole seething mass, feeding
through itself.

        They saw it,
        they knew it, those women,
        in their pelvic cradles.

2

The organism
feeding through   itself. O,
we have since become   so alone,
so separate. The joy
of root in earth and bee in blossom was
once in hand
in hand: the entire neighborhood—
man, woman, and child—threshed its grain
as a single crop. Hardly a thought
on it. The only way. Machines
had not yet come
between man and man
and woman
and child
and neighbor and neighborhood:

fingers ripped away
by the cornpicker,
we can no longer hold on
in the round dance,
have gone astray.

We used to cut the corn
one row at a time, then shock it.
Those bundles were big.
They hauled it down to the homes,
and then they'd stack it, and they'd husk it.
That's when we had those husking bees,
you know, and once in a while
we used to come across a red ear,
and the girls would help with the husking and,
if the girls got a red ear, the boys
would all kiss the girls.
There were never too many,
but there was always
a red ear or two.

Came from where, this custom
which fed
lip to lip,
corn to belly,
mystery. . .?
Came from what hidden
corner of the deep
mind, tradition, the deep
universe, carried
along through what migrations?

This custom,   pure expression
of that Woman,
like the redwing strutting, singing
through its mate toward
a circle of eggs. Same thing.
No thought,   just rising
to its tiptoes, singing, and the phallic
red ear brought bright
young life   lip to lip,
and the joy-song of husked corn
grew in the circle
of hands feeding the one
body.

3

Threshing time,   sacred time —
earth-Christmas!
The neighborhood became a body
with a hundred hands:

> *People come together at thrashin' time.*
> *O, it was a shebang!*

Earth turned   that golden form
which nourishes man
and woman and child, celebration
stirred in the blood and mind
like change in season:

When the steam rig came—
when they got to the edge
of the grove, the engineer
would start blowing the whistle,
because he knew all the kids loved
to hear it.

And the little rascals danced
and waved their arms
that such an outrageous sound
should bite the peaceful prairie air.

And meals!
Enough to kill yourself. O, we'd eat so much—
O, such good food— On the rig I was runnin',
Joe Kubesh's, south around Bechyn, those women
really put up a meal, Bohemian women, they were.
When they brung that lunch out, O, for cryin'—
Baskets of kolacky with prunes in them, donuts. . .

The men slept out in the barn,
'cause there wasn't room in the house.
Slept outside on a straw pile.
That's what we did. For sixty days.
Or a hay pile. Could hear the horses snorin'.
Slept just fine. Sure. Jeepers Cripes, yes!
Take a couple of blankets and cover yourself,
and you'd sleep like a hollow log
all night long. No trouble sleepin'.

Well, she was a great life. Yah, yah,

them thrashin' days is gone from me now.
I wish to God I could turn back to them.

Fed their lives   with that twist
on buffalo grass. O, a bit of cash,
but their bones and the glow in their eyes
rose straight from that seething mass
on which they walked:

In the fall, you'd take in your load
of wheat to get ground,
and you'd get a ton of flour.
There wasn't very much problem
getting rid of it.
When you had eight people
eating out of it,
about thirty loaves a week,
it went down. . . .

4

And work   found a way
to dance, the great
round, the organism alive
in shared movement of feet and hands.
Woman paired with man, the light
rising upward along the spine
to shine from eyes.
While the newborn, our grandfathers,
grandmothers, slept
in corners, whole neighborhoods

turned to the caller   inside
tiny houses, warmed boxes
in the vast
prairie night.

There used to be the old house parties,
and everybody'd get together, and they'd
have a good time. A square dance
would be eight couples.
We had a pretty good-sized house
for that time, a living room fourteen by sixteen.
O, that was a good place to dance,
An accordion, and sometimes a violin and banjo.
And everybody brought a cake and sandwiches
or something. It's too bad we can't do that
once in a while now. It is, by gol, yes.
You knew everybody there, and you danced
with everybody. . . .

5

But,
within it all, flexed
the desire, tough
and deep as the fiber
in a man's muscle,
to corner the prairie,
its ten thousand years
of stored life and death
in a man's house, his bank account,
to take it neat

as the steam rig's whistle,
down Allis-Chalmer's trough
into the mouth.

Good men and women, too.
They had a mission
which kept them sane
in their madness.
There was Henry:
pioneer yet in 1960.
I'd been hunting, heard:
Pock! *Pock!* *Pock!*
Slow rhythm of the axe:
mantra sounding through
wooded hills at the edge
of our prairie,
leading to odd moments
of calm and meaning, well being.
Pock! *Pock!* *Pock!*

Eighty-one that year.
He sat on another downed
cottonwood log
within the radiance
of his calm.

"Just this last small stand
of trees, and they'll be gone.
Then I can rest. Too much
brush in this country."

Around us the great
cottonwood trees, six feet
at the base, yellow leaves
rattling easily, as if they might
for yet another
hundred years.

His black and white dog, trembling,
sniffed my knee, darted back
from my hand.
The old man chuckled, but
his animal knew the better
who I was.

My voice sounded
from somewhere strange
and out of center:
"Aren't they paying
to keep land
out of production?"

As if we shared a joke,
laughter danced in the shallows
of his blue eyes.
"O, I don't know anything
about that. How can people live?
There's too much brush
in this country."

The black and white dog stared

after me, a pleading
in his eyes.

Through that afternoon, down
the harsh years
I've listened—odd moments,
eyes shut—to the secret
of Henry's calm:
Pock!   *Pock!*   *Pock!*

III. Hunters

1

The frontiersman, the hunter, had a hunger
to reap where he did not sow, to know,
as did the weasel, the fullness
of reaching into Her lap
for a round of sweet
blood-soaked fruit.
A sort of harshness grew, cold joy
in power over the creatures She expressed
in every nook and fold
of Her body.

We drew a bead on the land:

2

Late Novembers, surfaces
of prairie potholes and marshes
became a shield between
bitter air and earth heat

that life might scuttle
and unwind beneath, balancing,
and young men went out onto,
could see down through transparency,
could see bottom litter
of mud turtles and snappers
big around as Mother's washtub,
suspended for the winter in the wondrous hum,
could see, scattered
green blooms and fans
of the underwater garden, minnows and insects,
bits of the great   complexity.
All led to wonder, but
muskrat pelts were worth
a sound dime, and their sprinkled droppings,
light gray on black mud, marking
the winding way of muskrats through cattails:
push-up to bank den to feeder house.

The way
those 'rats travelled down there without air
was to breathe out against ice:
white bubbles over black water shimmering
like delicate lungs, breathing
oxygen from water.
(It all depended on that shimmering
white cluster of bubbles.)
And the 'rat
would breathe it all back
and continue her round.

O—root, tooth and blood!—
a young man with a big stick would leap
upon the top of a frozen house,
and *sploosh!*
the 'rat inside would be outside,
could be seen peddling through water,
strange blur of colors:
black skin tail and hind feet, back
brown and with an orangish hue.
And the young man with the big stick
would follow tight, great life brimming
in his blood, tightening his fist
on the big stick, and when the 'rat
in time must stop and release
the precious stream   of white bubbles against ice,
whack! whack! with the stick.
Drive her away
from her life.
Once, twice, and the third time—
out of great inner emptiness—
she'd jerk, toe to nose, in rapid fury
of need, then turn   belly-up.
Chop a hole, lift out your 'rat, whack
'er over the head, and "You've earned you a dime."

While one thousand white bubbles shriek
upward into winter's chilly sky.

Could take fifty in a day that way.
Then, in love with luxuriant fur,
you had your skinning and stretching

by lamplight,
fat and blood to your elbows,
late into the night.

3

Early winter winds, harsh,
out of the northwest, kept reaches
open on lakes, and that black water
(its unfinished ice formed in your flesh
as you looked at it)
would fill with late ducks:
mallards and bluebills, ringnecks and ruddies. . . .
Tough to hunt, to get at
across thin ice, but
some of the boys, amidst clicking of balls,
smoke and beer fumes at the pool hall,
devised a recipe: four quarts of whiskey
poured   into one bucket of corn.
Let 'er soak for a day or two. Stir
now and again and taste with a finger.
(Let a bit
of the infinite in.) Then
as close to open water
as a fool with a bucket
of whiskey-soaked corn dare go,
sprinkle your mixture
around on the ice.
Then, "git," and check back
every two hours, till

you see ducks floundering, drunk
as Hilding Quist, Saturday night midnight.

Keeping watch for the warden,
get out there and wring some necks, but
be care—
Jack and Ally each,
with a sack of mallards on his back,
caught up   in the blood
joy of the hunt, caught
in the sinews, puppet strings of the great
drama, great universal body which must   flux
through itself,   chased
a fluttering, quacking mallard as if it were more
than a meal, a sort of grail
led to thin ice,
and they went down through
whirling in loon shit.
Long streams
of white bubbles from lungs
and marsh gas released at last
from deep rot. Then,
nothing left
but tracks,
duck blood in snow,
mud-rimmed hole, moan
of the northwest wind. . . .

Two men hugging ducks,
feeding roots
at the bottom
of Goose Lake.

4

Went north—they still do—
for the deer hunt.
Early November man-rite. Bucks in rut.
The blood-colored wave washing
up from these prairies. Even then,
back in the thirties, the primal sense,
the joyous artery which throbs through
everything,
had to be hunted.

Paxton's Cafe in Olivia—
lunch counter with a bar at one end—
served them in passing:
splash a bit of the burning amber
across the tongue, warm to talk
of old hunts; cold stands,
hour after hour, shot deer crashing
through thickets, blood in snow,
the deep welling
of a strange love. . . .

Arlen remembered them,
four brothers, red-coated, from down near
the Iowa border. On their way up.
Rowdy, with extra
whiskey   to warm the road.
O, they wanted deer so bad
it ached in the curve of their finger bones;
they could just taste   the gall
in a bit of liver, raw.

"It was 'bout a week later.
Three of them come back,
grizzled with beard, all red-faced and tired
as a pack of twenty-year-old hounds.
Drinkin'. They was drinkin' steady,
and no talkin', and no laughin'.
I got a funny feelin'
they didn't wanna leave. You know
how drunks are ashamed
to go home to their women . . .?

"Finally, I asked, 'Well, how was the huntin'?
Wasn't there another one of ya?'

"They'd shot 'im.
Shot their own brother
at the crack
of that very dawn.
Squatting
behind a bush, takin'
his mornin' constitutional.
It was the white of his ass
and the toilet paper.
He'd waved it, and they took it
for the tail of a buck.
Two of them cut loose,
emptied their rifles
through that bush.
All excited, I guess.
They ran and found
their brother, Eddie, it was, kickin' brush
with his pants down. All shit and blood.

"Died in less than ten minutes.
They said he kept screamin'.
You know,
they had him out there, outside the cafe,
wrapped in blankets and strapped
'cross the top of their car,
alongside a goddamned deer.
Said they couldn't stand
to have him sittin'
'long beside them
all the way home."

5

Clean up the varmints!
The buffalo, the cranes and passenger pigeons. . . .
Hell, we're still potting away.

Tom's dog, Pintail.
Part Chesapeake, we thought. Great
retriever in his day, but old now,
never bothered
by veterinarians, their needles and knives,
and finally the manure spreader
swung out and around faster
than Pin's waning reflexes.
Had to haul his right hind leg
ever after.

Last hunt.
He sat down, green ooze to his neck,
at the center   of a square mile of cattails,
acceptant, willing to sink down and return
the white stream of his life,
give his flesh over
to snappers and blood suckers, except
Tom and I, sixteen, hip-booted
and laden with mallards,
somehow still living near
the prairie's beginning,
carried Pin in our arms to a wide and deep ditch
at the swamp's edge.

We rested while a black spot
grew large in the western sky.
I snapped off a shot, and it dropped, *splat!*—
spoonbill duck—at the far side of the ditch.

"Fetch! Fetch, Pin! Fetch!"

We used sticks, let Pin smell them,
tossed them in the direction,
hoping to prime that primal
energy from the sky
through us into him.
A boost with the foot.
Pin barely raised his head, till
a stick hit the spoonbill and released
a final flurry of life
through wings.

As if to catch it for his own,
Pin began
slowly
swimming
toward that splashing,
fought the heavy moss
and scooped
the spoonbill up in his jaws,
continued on to the far shore where
he sat down and ate the duck,

death or whatever. Pin,
with his thousand years
of good breeding, was soon gone.

Tears in the eyes of young hunters
brighten the edges
of the black spot which hovers
in our western sky.

IV. As Within, So Without

1

Dirt was cheap, and greed
burned behind keen eyes of leaders
of strange new hunting groups:
moneyfamilies and corporations which sucked
wealth from earth—
sunlight and angleworms transformed—
faster than it could be returned.

They could beat more life
from Her body in one hour
than a young man with a big stick
in his lifetime.
They called themselves the business
of America.

2

Hard times, and hard-used
human flesh flowed into new vacuums, free-loading

along those steel lines which cross-hatched
the land, funneling maggoty carcasses
gone wheat and gold, into the maws
of still-hunters at home.
Hoboes, often beaten, but needed
in that system, rode the rails
and shocked the grain.
Men who couldn't fit in or measure up
to the struggle (such was the talk through cigar smoke
in cities, where Gentleman Farmer Clubs
met for lunch on Tuesdays
to discuss the price of grain), but
hoboes brought the harvest in
cheap.

Hoboes used to lie
under a tree
in Sleepy Eye, Minnesota,
and farmers fetched them
as needed, but,
at the peak of harvest, woe
to the man whose voice had raked their backs,
whose wife had fed them mutton
instead of beef:

> "How many acres of oats
> you got down out there, Jarvis?"
>
> "Something over 300."
>
> "Well, you bring 'em in here to town,
> and we'll shock 'em for you."

The materialist, reaching
for Her lap, Earth's basket,
which should, by God, oughta come
when you have hard-handed wrung
the juice and blood
from your land and family,
every moment of your day
and life,
finds Her
vanished,
the soil dry
and drifting,
the hoboes laughing. . . .

3

As within, so without. Barren
black deserts of our fall-plowed prairies
are an end to wonder. They shut out
the light. Life
grows dull.

"Three pounds of beef
makes a good stew.
That's my philosophy."

He's dead now.
Was nearly dead then, except when lit
with Three Feathers nipped
from half-pint bottles tossed

into ditches, and he sang songs full
of hard-used woman-flesh: "O,
the wonderful Winnipeg whore. . . ."
I remember that evening, accepting
his cold gift of wisdom, given
in hate of poetry, how
a wren sang a song full of twilight
in a lilac bush.

4

As within, so without:
once an aura
of bright songbirds—tanangers, red,
and orange and black orioles—spiralled
out and around the body.
Have gone black
as exposed soil.
Our birds have turned
with the sod,   turned black—
starlings and grackles waddle and croak.
The complexity of joyous song reduced
to survival: each half hour
one chirp of the sparrow.
Barely enough, barely enough
to turn a planet,
to raise the sun of a morning.
Sooty birds of civilization—
sparrows, starlings, grackles—
have become our aura.

Across the great black belt
of corn   meadowlarks now
sing only in ditches.

V. Black Wind

1

This vast
prairie, its hide of sod
stripped back, black
living flesh of earth
exposed.

Our way
has made thieves
of the wind and the rain.

Listen,
listen to the wind moan
through the bone-white dead
cottonwood limb: *Half gone!*

Half gone! *Half gone!*

"Where? Where's it hid?
Hell,
I'm just swappin'
some land
with my neighbor."

But see! Over there:
That fencerow ridge
with its line of sod   up
a foot above
its field:
Half gone!   *Half gone!*
        Half gone!

And see!
That little graveyard:
stones, poems, a crucifix. . . .
Built on a knoll
overlooking the land.
The old ones lie there.
Those who tamed, turned
it all over.

"Built on a knoll, you say?"

"They all were.
The custom back then."

"No, no that was no knoll.
That's the only spot
untouched by the plow.
The rest has gone away."

Gone away
like a childhood game.
"Pom-pom pull-away,
you don't come,
I'll pull you away!"
Gone away
like a dead mother.

Our old ones,
our great grandparents, hold
last virgin earth within
their thicket of bones.
Despite "Rest" and "Peace" engraved
on each stone, the elements have written
harsh words: this slight rise,
its surrounding vista.

2

The great herds, the old
rivers of flesh—
antelope and buffalo—
all plowed under.
And the wind
is blowing
them away. . . .
Press your ear to the ground.
Hear the thunder? Your mind
can hear it, great river of life
down in that darkness

still flowing!
Could rise again!

"It wasn't more'n ten years back
conservation meant deer
and pheasants
to hunt."
Down now to the bare
grit itself.
A handful of dirt,
handful of thunder, swallowed
by the wind.

3

I think it was about '33 or '4
when the eave spouts in town here
would get plugged up with dust.
It would come into the houses
right through the window casings.
Talk about a mess.
It was just any place you'd take your hand,
you'd get it full of dust.

It was so dark
the chickens would go to roost
in the middle of the day.
Awesome
loss of light.
The sun eclipsed
by stored life.

My God,
you could be out,
and that wind would start blowing that dust,
and it would get dark as it is at night.
You couldn't walk agin' it at all.
Well, I went over to my father-in-law's place.
I'd taken the missis over there,
and he had a fella working for him
by the name of John Lane.
And John was goin' out to pull rocks.
I says, "John, I'll go out and help you."
And he was the damnedest fellow.
He wore a high-top shoe,
but he'd never lace it up.
Let that string drag along behind him.
And we went to load up a bunch of rocks,
and the fella says, "Stanley,
I believe it's going to rain."
I says, "Rain, hell,
that's just dust you see a-comin'."
We loaded up the rock and unloaded them,
and went out to where we'd left
our pry bar and spade and shovel.
And John says, "I don't believe there's any use
to loadin' up another load. It's goin' to rain."
I says, "That's not rain, that's dust."
Well, just about then the storm hit us.
And John says, "I'll get the shovel."
And I seen his shoestring layin' there,
and I stepped on it.

John was down, and he says, "By God,
I've got to tie up my shoes."

And boy, we couldn't see the house!
We was maybe a quarter
of a mile from the house.
We couldn't see where the place was.
But them damn horses, they knew where to go,
and we started them out.
We rode on the stone boat,
and we come right to the house,
and Marion's father come out,
and he says, "What's the matter?"
"Nothing," I says, "only it got so black
we couldn't see where we was."
I was standing on John's shoestrings,
and he was falling around there
like he was getting blowed over.
And you know, it was just so dark,
and the fencerows, they was blowed in
with dirt and weeds. Anything
that dust could catch on, it would stop on that.
O, God,
they lost a lot of topsoil then.

Gone away, gone slick down ditches,
down to the gulf.
The stuff of new life, hidden
in places we can't get at.
Four billion tons of Her prime,
year in and year out.

O, we peeled back her hide
and let her blow, till the rooster crowed quit
in midday, but we could not, would not, cannot quit,
and She's drifting away faster even than then! O,
the old team
has been shot and fed to mink,
can't bring us home.

4

Thin, thin, seething layer—
six inches, a foot, none on the hilltop—
the sum total   of past life
on these prairies, the end
and the possibility   rising through
again and again:
bee, buffalo, bastard grass and man, all
have rotted in and risen
again and again.   Aeons
of struggle, blood and root,
all come to this   black stuff.
And we—
the life of this happenstance moment—
are a haze shimmering above.
It has all come to this
black heaven,
and the wind,
black wind!
            is blowing
                        it away.

VI. New Tide

1

White frame houses with red barns and sheds,
four and more to the section, sprang
upward with that same
power of first wheat.
Four to the section, on the grid.
Then the little schoolhouses:
dark winter dronings
over decimals and fractions,
pledging allegiance
to that way. And the grid filling
with churches and small
outpost towns with mainstreet stores,
served by spurs
of the great railroads. It all
blossomed on that grid,
as if forever, for about one life:
human community, neighbor
and neighborhood, small farm, small town,
everyone dancing

with everyone else
on Saturday night. Then

the thin intense stem
of American technology   blossomed and bore. . . .
It all began
to pass. People left the land.
The white stream again released, replaced
by machines,
chemicals,
profits
somewhere else.

Small white schoolhouses and churches
now store straw. Half the houses
hauled to town or burned to the ground.
Space created   that machines
might spread their wings.

2

Machines.
Remaining men race   machines
over the land. Eighty thousand bushels of corn,
immense golden cob, swinging
out ahead: wish-dream
of luxury, sensual ease, time
to go hunting.
Ever receding dream. So hard
to gain a true hold with the finger bones.

And machines rust, petroleum and dollars flow.
O, those four-wheel-drive tractors multiply
like the jackrabbits which once
romped jauntily over these plow deserts.

"These days, yah, there's a lot
of recreational tillage."

Receding cob of gold,
and the red ear has vanished
in hybrid magic, while
for one moment, we enjoy
something like eagle flight:
the power of machines
turns earth to air.

3

"My dad said something
about nitrogen fertilizer.
Said, 'You take your hired man
to the local bar some morning,
and you'll get a whole day's work out of him
that afternoon,
and some singin' besides,
but what about the next day?'

He said, 'You can take a little kid
and inject him
with some of this fancy bee's wax,

and he'll grow
to nine feet tall,
but what kind of kid is that?' "

4

Out south of town
any town out here,
the old tree claim,   planted
by first settlers
for firewood, where people picnicked
and kids built forts in thickets,
played Indian, hunting the buffalo
which still live in the deep
mind of children, of us all, waiting
for that alchemical moment
when what is base
is transformed
and once again the great herds
emerge with the grass.

The old tree claim,
an oasis for people
in endless clean
agriculture,   has become
a chemical dump,
a place to drain away
last flickerings of liquid fire
from the tank.

"What in hell else
can you do with it?
The rest is under
the plow. Land's *gotta* be farmed."

Only the tree claim, the old grove,
ten acres covered with bare,
dead spots. Nothing
will grow there
for years and years. Strange
tough tiny plants begin,
trying to heal
the burn—poor scabs
wane away.
Trees dying. Children
with bows and knapsacks, lunches packed,
still play in thickets, dreaming
of conquering
wilderness.

5

In that last cold month
of '77,
we began to seek
pure water from springs.
It came to that.

Withering plants in window boxes spoke
of strange slow fire flowing

from water taps.
Like canaries in gas, these plants,
their green tongues spotted with brown,
spoke.   It came to that
in the last cold month of '77.
We sought springs.

These thousands of square miles of blackened prairie,
our home, once a skein
of ponds, creeks, streams
lighting the eyes of endless migrating geese
(but one lifetime back!),
now
only these few square feet
here and there, hidden, soft openings,
lipped by living green, clean flow
of surface water fit to drink.

Clean, clear
of the corporate burn,
slow burn to the bones, the nerves, nourishing
sweet flesh blossoms
of cancer.

It came to that.
We knelt to springs,
while around us tranquilized cattle
to their hocks in feedlot mire,
munched strange grain,
and strange children
with stress rings 'round their eyes

stared on. We knelt

beside a thousand years
of pure flow, washed
through the mesh of deep earth.
This light, this grace   up
from blackness.
Sought now, rare and precious
as glittering stones.
We knew this in our anxious bones.
As others, in a different time, sought
pure action, pure thought,
we sought pure water.
Something has been lost.
Something has been sold away.

We kneel now beside a thousand years of grace,
plastic bottles in our hands.
Our tongues touch
the pale fresh flesh
of pure water.

6

"Soon won't be enough
farmers left in the U.S.
to fill the Rose Bowl."

"One hundred thousand super farmers
is all we need."

O, to be one of these!
Even now
one man in seventy-six
works the land. The rest have been freed, not
to sing hymns, bless life and bountiful nature, but
to produce
TV
dinners,   automobiles and blow
dryers. A pyramid of stability flipped
onto its tip. That's the point:
one man in seventy-six
works the land.
Backed by the big
tractor and chemical tide: a new sea rolling
over beaches which once held
Lake Agassiz.

One man in seventy-six, master
of the feedlot, cattle by the hundreds
to the hocks in chemical mire,
tranquilized to bear
long days of mud, shit, and feed
calculated
to explode cells.

One man in seventy-six juggles
the vast array
of our food machine:
debeaked chickens, "food things"
in skyscrapers of cages
where they wait   laying

day by day   unfertilized
eggs juggled to keep
red from black.

Juggling
the feed lot, the chicken ranch,
the conveyor belt tomato, built
by man, by God, tough and four-square
to withstand the rigors   of machines
and teeth.

One man
in seventy-six. It's become
high stress, ulcers and the heart
tearing at its cage of ribs.
The tranquilized farmer marches
through technological mire,
past dreamy-eyed cattle,
to push the button
which is his day.

7

Corporateland sends out a man
to sell   its
vision of the future:

"Farmer Brown," says the man from Corporateland,
"will have a machine
the likes of which you've never dreamed,

but *is* being tested
right now. It'll do it all.
You won't plant taters. You won't plant cotton.
You won't plant corn and beans.
You'll plant *cookies.* Everything you need
to make cookies. And the cookie machine's
got a brain, and will roam your land,
tilling the soil with microwaves, and
spitting
intellectual capsules which will feed
your soil everything it needs,
as it needs it, and come cookie-time,
they'll all be wrapped in paper made
from leaves and waste
of plants."

Farmer Brown's not needed
in this plan.

Come cookie-time.

VII. Bullish Americana

1

We used Her, chewed Her,
unfeeling, to the bone,
and when the good was gone,
we spat out the gristle
and moved on.
Always a new green hand, beckoning
from somewhere other.
Some new, trim
virgin body in exact form
of our desire.
Load up the wagon and begin again,
fresh. Life will do a dance there
it never did here.
Yes, when the wheat rose
weak   and turned
yellow, we turned
our backs. We moved on.

2

Our passion for land
                                        wanes
with the bloom. Blood
of a torn hymen (proof of power)
is needed
to keep us gentle and attentive.
O, the plow turned the virgin
prairie
over, and her youth flew,
with the wind, and the rain
carved unlovely lines
in her flesh.   *Half gone!*

Our passion wanes. We move on.

"Hell, it's human nature. It's like
when you're married. You take
a pair of newly-weds,
and put a bean in a jar
every time they do it
the first year,
and take a bean out
every time ever after,
and you'll never empty that jar."

Our passion
                        wanes.
Hell, that half-gone woman
is still good. She is still good.

Will reward love
with more than we could
ever dream.
One seed in her belly, cared for,
unfurls, in time,
into trunk, leaf and limb.
She is still good, all
we will ever have,
will ever need.

3

The eunuch served
and sang.
The Hercules god-man
was sprung at the hip
to limit flight.
The gelding is gentle
and serviceable.
Beefeaters find
the steer to their liking.

I am laughing a bit in my chops, but
there is a kind of castration
which is needed
if we will survive
the wound (it has become such)
of being male.

4

Went to court: Swanson vs. Johnson.

Johnson's lean and somewhat aged
Holstein bull, McCarthy,
acting out that role which pours
from bull through cow
to bull,
had walked through his fence,
down the road and into
(right here in Minnesota) the promised land:
a pasture of heifers. Forty-two,
sweet with sleek furry flanks, their flesh
somehow now a hollow
moaning
for something needed from
the opposite pole.
And they accepted
that black and white deliverance
which, though aged, could still make it up
onto its hinders, and rock a bit,
and a new beginning was ignited
in the flesh of each.
They were at peace, but

Swanson vs. Johnson.
Swanson's thought:
"That lean, spavined, and mean
creature, Johnson's bull, couldn't produce
a calf worth the half its feed."

And all the while
the Mother was smiling
langorously.
In Her economy,
ingenuity and singleness of purpose
often come to reward.
That She should experience such
sensual rapture in Her multitudinous flesh
was enough. Forget the courts.

Swanson, of course, had planned
on the man with the long rubber glove
(always called Ferdinand)
and his tube of cold semen
milked from a prize
Danish bull in its stall
in California.

Well,
the jury hung, and Johnson
sent Swanson a load of hay.
McCarthy made the papers. Saved himself
from hamburger for a season,
became a sort of regional showpiece,
up to his hocks in daisies,
slobbering green grass, the Mother
serenading his days
through the throats
of meadowlarks and bobolinks.

5

There is a great sadness:
the things a man can do,
unneeded,
in a new age.

"I used to hunt, but. . . ."

I remember a white-haired old man
working fish nets at freeze-up,
his heavy hands dipping
into bitter water, pain set aside,
into some compartment, not to be opened
for a time. I remember another
powerful old farmer who found me far back
on a two-track road through brush country,
stalled in a blizzard.
Hauled me for hours.
(So many times stuck!)
His shovel, his shoulder, his pickup,
canny lore of weather, roads, and machines
kept us going, and in the tiny town of Fosston,
he shoved me—frozen package delivered—
into a garage, said:
"Here's a young man down
on his luck." Then was gone.
Couldn't find him. Couldn't pay him a dime.
Didn't know his name. Had he ever been?

"I used to hunt, but, hell,
the gun's gone so rusty
it won't even shoot no more."

Great sadness,
such things that men can do!
now so seldom needed, and yet
a new interior, weeping
for heroes,
yawns dangerously
before us. . . .

VIII. The Houses of Lovely Women

1

The houses of lovely women
in their middle years and late
are often a refuge,
last refuge, for the game
of light. . . .
Long shafts of sun reaching
through clean panes to kiss
green leaves of plants
(all here is kissed
in the mind
if not the flesh)
and flashing
from frames of photographs, the loved,
once nurtured within these walls,
their toys—stuffed animals
and basketballs of the past—
now set in darkness, kept to light
memory of other times and love
which continues
as sounds and smells from Thanksgiving

kitchens: the good
silver and china (bone perhaps, Grandmother's
remnant of the end
of the great herds) conveying
tried recipes of the past.

Ecological balance within walls,
within the vast   wilderness
of the heart,
where more is given
than taken away.
Last refuge, a place
where dogs don't
eat dogs, not
on those rugs, scraps
of flesh, hair and bones.

But, as to how
those houses were bought
and paid for—
Draw the curtain!
Close the door!

2

Household balance. Lovely women expressing
inner nature,   effortlessly,
know how to touch
the world. Yes,
that violet placed there

amidst that lace and light, yes,
and that bone-white vase
by itself on that shelf
over the buffet. . . . A way
found at the inner center
of women, a true expression
of nature, as natural for us
as anthills for ants,
beehives for bees, oriole's nests
swinging over water,
woodchucks tunneling under
to spend winter
in the ground. . . .

We must learn to live
within the great house with the grace
and balance lovely women have maintained
in their homes in the face
of (and supported by)
the bulldozer,
dynamite, herbicides, and endless
ramifications of multi-billion dollar
multi-national corporations, arranging
land to their liking, spitting
bright birds and soil life—
insects, nematodes, bacteria, millions
to the teaspoon—
aside as they ride upward toward
profits   unrelated   to human welfare
and that of the great
house, the whole

which is our home. Ah,
learn to touch
like a mother.

3

There's an old Japanese—Masanobu Fukuoka—
who, when young, saw
that Man can accomplish
nothing,   saw
that nature carries us, we ride
Her waves   or
drown.

Masanobu sought to express
what he had seen
through the small farm of his father, ruined
a good, well-pruned orchard of mandarin oranges
before he saw
that to go back to the Way
requires patience and gentle guidance.
Those trees still knew
how to grow
toward health   and abundance,
but their training
had sent them askew.

Now, these years later,
Masanobu's fields grin with the sun,
crop after crop,

never tilled,
one flowing through
roots of the other—
rice, barley and rye—
season by season, easily
as grasses of the virgin prairie.

Masanobu does not till, he touches
lightly, and with wisdom,
and he harvests.

IX. Awakening

1

They were tough, had to be,
those who swept it clean and took
a million
square miles of sod
and tipped it all over.
Forget greed and lack
of compassion, consider a vision
of new light (do not try to deny
its evil glintings) sweeping westward
from European misery.
If it had to be done,
hard men and women
had to do it.   Maybe.

If the greater body wants. . .
new mind, new light—
O, the bright faces
of our species! Just
as your child's eyes light

your days,   new eyes, new minds
might— Antennae emerging
from the heart.   If
what is great   wants
new light   or that old
of those who once lived
on the balance, but
more abundant,
that light   amplified evenly
over its entire surface
that it might joy   in itself
the more, know
its own joy, even guide
its growth and change,
hard men and women
had to begin it.   Maybe.

Perhaps, then, this long-lifetime loss
of balance, this time of snout and jowl,
is preparation for THE GREAT—

2

The light shines through
each blade of prairie grass,
each mite, mole, mushroom,
running killdeer
is wondrous. And the human face
when lit
will be a beacon

for galactic travel.
This is not hyperbole.
The light shines through!

We belch over pork,
worry about weeds

while we drift amidst galaxies,
God-light,
infinite song of the infinite universe. . . .

3

All that life   has now been turned
human flesh, up from those wings
of birds and insects, up from
all those songs, dozings in the sun
of deer and foxes,
all the long hours of animals tranquil in twilight,
all that life of the prairie, swept clean
to black grit, blowing,
all that life, all that death
sleeping in *our* flesh.

We owe so much.

4

We begin
to become real   only

when there is no other way,
when we stand face to face with our death,
can see
the rolled back whites of its eyes.
We begin then to correct   our lives,
the ecology of our days   and pray
it's not too late.
Now that the cancer
has blossomed, that the snowstorm
has begun in the blood,
now that the heart tears
at its cage of ribs,
now we will begin
to pray:
"Only show us the way."

We have now arrived
at the dark spot, are in it.
Not knowing
we've been gnawing our own   entrails.
Somehow hoping never to see
the lidless corpse, hoping
to stuff and cap the appetite forever
and bank
immortality. Such fantasies!
Christ-God-Buddha, wherever
enlightenment lives, let us feel
our pain   to the bone
that we might begin
to correct our days,
and pray

we are not too late
and give thanks
that we suffer
in so many ways that we must
soon begin, as if we meant it, by God,
to sort out, to seek and find
the holy balance.
What a privilege
to be born into the age
which will have to suffer
toward the great   leap.

5

There could be, there might be
a global neighborhood, a body
with ten billion hands, and
scales, limbs, leaves, all
else intermingling, knowing. . . ,
and people who need   less and less,
their machines simple or simply
forces of the whole   guided
gently toward the dreamed garden—
those plants aquiver with eternal
magic—as ever.

There may well be people
who seldom leave the region
of their planting,   gardens
of collaboration; the forces

of the organism breathing through their sensitive
fingers and minds. A continuous agriculture—
like currents at the bottoms of oceans—
where the red ear of corn
always grows.

People who seldom leave
beauty and sustenance   continuous
with their flesh and mind, but,
with their eyes closed, the whole
is with them   always,
and they are full   to their fingertips.
Who, with eyes closed, beam outwards and explore
the planets of Andromeda.

Who need only sit quietly in twilight
chanting, gently, some loved word
for endless miles of nerves
to ignite with the immense
symphony.

6

TIME.
        IT'S TIME.
Time now for it,
the great organism
that we are,
to begin
to stir and stretch.

Time now for the great dawning
in its mind, our minds. Do we know
our feathers from our foot, roots
from fin, tail,
smooth expanse of skin. . .?
Do we know
that our hawks
send our rabbits to sleep within
our own breast?
To awaken again
as grass? And again flesh?
Thunder in our flesh!
Do we know
that we all sing the immense
bird song which lifts the sun
each dawning?

All sleeping now
in our flesh, beginning now
to stir and stretch.

A new dawn, another new dawn,
new age in which we know all
that's sleeping in our flesh—
I don't think it can be stopped—
must begin its morning rage
of song.
We must make it sing.

Following 15 years of "alternative living," in both urban and wilderness settings, Joe Paddock in 1975 became involved in a series of rural arts and humanities residencies: Writer in Residence for Lakewood Community College in White Bear Lake, Minn.; Community Poet for the small town of Olivia, Minn.; Regional Poet for the Southwest Minnesota Arts and Humanities Council; Poet in Residence for Minnesota Public Radio (at Worthington); and Humanist in Residence for the American Farm Project of the National Farmers Union. He is currently a humanities consultant with the St. Paul based Land Stewardship Project. He has published many poems, stories and articles in magazines; has had two chapbooks (*Stored Light* and *A Song Like My Own*) published by the Southwest Minnesota Arts and Humanities Council, has edited or co-edited more than a dozen project books and booklets, and has published a book of oral history, *The Things We Know Best.* He currently lives with his wife, poet Nancy Paddock in Litchfield, MN.

Photo by Tom Guttormsson

For my daughter, Erika, who is a constant source of inspiration to me.

ACKNOWLEDGMENTS

This book would not have been possible
without the encouragement of
Philippa Caldwell Motley,
Janet Gluckman-Berliner,
and
Shelley Adina Bates.
My love and thanks to you all.

CHAPTER 1

Erin glanced up from the message she was texting just as her uncle turned off the main highway. An enormous sign stood at the side of the road, which read:

POSITIVELY NO:

1. Photography
2. Tape recording
3. Sketching or drawing
4. Picnics
5. Camping
6. Fishing
7. Swimming
8. Removal of ANY objects
9. Alcoholic beverages

VIOLATORS WILL BE PROSECUTED

Erin lifted her bulky backpack up off the floor of the ancient pickup truck and cradled it in her lap. She could feel the contour of her camera pressing against her side. *I'll have to hide it*, she thought uneasily. *I wonder if they'd take it away from me if they saw it. I wonder what they'd do to me if I took a picture with my phone.*

"We're almost there," Antonio said, startling her.

"T 2 U L," she sent and hung up. After calling her parents to tell them she'd arrived safely, she'd been texting friends nearly constantly since her plane had arrived two hours before.

She looked at her uncle Tony, examining him as closely as possible without being conspicuous. He'd hardly spoken on the long drive from the airport where he'd met her. He looked so serious, his eyes fixed on the road ahead. He hadn't been in the least quiet the past summer when he and Aunt Julia and Erin's cousins had visited California. He even looked different here in the desert of the southwest, somehow mysteriously darker than in the living room of Erin's California home.

He was as handsome as she remembered, his straight black hair shining in the sun like melted crayon. Here, there was something about him, an intensity that made her uncomfortable. Had she made a mistake, coming here for the entire summer? She'd just graduated from high school and could have spent vacation with all her friends getting ready for their first year of college. Instead, she was keeping a promise made to her cousin the year before.

Two months, she thought. *That's a long time to be away from the family.*

But he is *family*, Erin argued silently with herself. *He's married to Daddy's sister.*

"There's the dam," Antonio said, pointing to a huge, curved, concrete wall that obliterated the entire view to the north. Water flowing down the spillway sparkled brightly in the afternoon sun.

Erin noticed that in the mile they'd driven since turning west off the major thoroughfare, the desert that had stretched out on both sides of the freeway on the long drive from

Albuquerque had given way to the beauty of the Rio Grande Valley. She turned to look back at the river that spanned the length of the valley and saw that it was lined with lush green trees, their leaves shining as if an artist had brushed green paint across a dirt-brown canvas.

The far side of the river was a giant patchwork of light and dark green crops, planted in straight rows separated by irrigation ditches. Beyond that was another waterway, lined with the same trees as those along the river.

"Our water comes from the lake above the dam," Erin's uncle told her. "The pueblo is there, just beyond the canal," he said, indicating the direction with a nod.

His voice was gentle and soothing and drew her in like the pied piper. *He really is handsome. I wonder when Aunt Julia actually,* actually *fell in love with him,* Erin mused, still clutching the forbidden camera and paying as much attention to her handsome uncle as to the surroundings he was showing her. Erin's aunt had grown up in Albuquerque. Now the Indian Pueblo where she'd been teaching when she'd met Tony was her home. Julia and Tony had two children: Neyse, who was her age, and Carlos, who was fourteen.

She and Neyse were practically twins; Neyse was just one day older. The girls had even teased about changing Erin's name to Daisy so it would rhyme with Neyse. They'd had so much fun together in California; it was as if they'd lived near each other their entire lives. Erin had never visited New Mexico before; other than sleepovers, she'd never stayed away from home at all except at summer camp.

Maybe it won't be as much fun here as it was in California.

She glanced at her uncle. *Maybe he doesn't like me: he's hardly said a thing.*

Tony drove across the bridges that spanned the river and the canal without slowing down. Normally, Erin would have asked him to stop so that she could take a closer look at the water, but she just held her backpack and didn't say anything. She didn't want to be bothersome. He turned onto a hard-packed dirt road and then, rather quickly, into the driveway of a house enclosed by a wire fence. The yard looked as barren as the land immediately surrounding it—no trees, no plants, no grass, just dirt.

"Here we are," Tony said, stopping in front of the house. Jumping out, he came around to open the truck door for her. "Let me get your backpack."

She jerked it out of his reach. "I've got it," she said quickly, hoping he hadn't noticed her guilty discomfort. The edge of the camera bit into her side as she squeezed the heavy pack. Tony got her suitcases out of the bed of the truck and led her inside.

It wasn't at all what Erin had expected. She'd seen pictures of adobe houses, squat and sturdy, looking almost as if they'd grown right up out of the ground. She had assumed that since her family lived on a pueblo, they'd live in an adobe house. But the house she entered was a simple tract-type house, like hundreds she'd seen in California.

It certainly didn't seem very "Indian" and, somehow, though she was comforted by the familiarity, she was also disappointed. Her father had told her that the Pueblo Indians had used adobe for hundreds of years to build their homes and she wondered why Tony and Julia's house was not made in the ancient style. She considered asking him but his reserved demeanor made her decide against it.

"Why don't we leave your bags in Neyse's room," Tony said. "She should be home around dinnertime and she can show you where you can put your things."

Erin nodded and followed him down a narrow hallway to a small, bright room, with two bunk beds, pale blue walls and lacy white curtains. Posters of rock groups covered every possible space on the walls. Erin walked over to the delicate white desk that stood under the window. The top was clean, except for a small, lop-sided lamp and an envelope with her name on it.

Her uncle put her two suitcases at the end of the bunk beds. “Looks like Neyse left you a note,” he said.

Erin picked up the envelope and turned it over in her hand. She had been upset that her cousin hadn’t been at the airport to greet her. Now she wasn’t here either. Tony had explained the reason but Erin was disappointed nonetheless.

“Would you like something to drink? Juice maybe?”

Erin nodded. She didn’t really want any juice; she just wanted her uncle to leave so that she could open the note from her cousin.

“Yes, thank you, that sounds good.”

“—in the kitchen,” her uncle said.

As soon as her uncle had left the room, Erin tore open the envelope. The message was short:

“Hi. Sorry I couldn’t come to the airport but Mom needed me to help with her class on a field trip. We should be home by about 5:00. Help yourself to whatever you need. Top bunk’s yours, if that’s okay. Can’t wait to see you.

Neyse.”

Feeling a bit better, Erin put the note back on the desk.

“There’s juice here for you,” Tony called out. His voice came from somewhere across the hall from Neyse’s room.

“Coming,” Erin said, feeling thirsty after all.

She started to leave the room when she realized that she had to find a place to hide her camera. She looked around. It was entirely too neat to hide anything anywhere. Opening the closet door quietly, Erin peeked inside. On the left, several oddly shaped boxes were stacked on top of each other. Figuring that one more odd shape would never be noticed, Erin put her backpack with the illicit camera behind one of the largest boxes. *I’ll decide what to do with it later*, she thought, quietly closing the closet door.

“Did you get lost?” Tony called out.

Erin took one more quick look around Neyse’s room.

With the exception of her bags at the end of the bed and the now opened note, everything looked exactly as it had when she'd walked in.

"I'm coming."

She crossed the hall and entered the kitchen. Tony handed her a glass of juice and offered her a chair at the dining table. They sat together, drinking juice while Erin looked around the room. White walls and brown carpet blended into the brown and beige plaid of a couch that faced the fireplace.

Except for the piano that sat next to the couch, she might have been in a dentist's waiting room. Couch, chairs, piano, fireplace; even the well-worn rocking chair looked generic, the way it occupied space under the two corner windows and was surrounded by jumbled stacks of books and magazines.

Once again, Erin was let down that the house didn't seem very "Indian."

Did you expect a teepee, for heaven's sake? she chided herself silently.

"Are you hungry?" Tony asked.

Erin turned and smiled at him, relieved that he had interrupted her thoughts. She wanted to talk to him, but didn't know what to say. "Not really," she confessed. "I had a sandwich on the plane."

"Airplane food would ruin anybody's appetite," he said with a soft laugh, and for the first time, he sounded like the man she'd met the summer before. "Well, then, what about taking a walk around the pueblo. I'd like you to see our village. Julia and the kids won't be back for at least a couple of hours."

"Okay. I'd like to change my shoes first, though," she said, looking down at her new sandals with a frown.

He nodded.

Back in Neyse's room, Erin resisted the impulse to check the closet to make sure her backpack was still there.

She and her uncle were the only two people in the house; who could possibly have moved it? Opening the larger of her two suitcases, she took out a pair of tennis shoes and socks. She'd expected the weather to be much warmer than it was, not just because it was the desert, but because her father had spoken about the terrible heat at the pueblo. It didn't seem any hotter here than it had in California.

She kicked off her new sandals and, after changing, she brushed her long, fine, honey-gold hair into a ponytail. She looked at herself in the mirror on the back of Neyse's door.

She was tall and broad shouldered. Her skin was fair, her cheeks were highly colored, and her nose was splattered with freckles. She had a golden tan most of the year since she spent so much time at the pool. Her wide-set green eyes had a slight upward tilt, something like a cat's, she thought with a frown. The overall effect wasn't bad, though, and she gave herself a thumbs-up sign.

She pulled out her phone and typed out a quick message to her best friend, Taylor. "Not so sure about this whole thing. Might have been a big mistake ☹" Then, she joined her uncle in the front room.

"Ready?" Tony asked.

"Yes," she said with a broad smile.

They walked toward the canal, where Erin saw a cluster of adobe-colored buildings. As they got closer, Erin realized they were houses, set both apart and close together in a random way. Some of them were made of concrete block, like Tony's, some were wood, and others were actually adobe. With the exception of a few wispy trees obviously struggling to survive, the yards had no greenery. There wasn't a single lawn, a dramatic contrast from the lush California landscapes she knew so well.

Dark-haired children played in the yards and on the road, riding bikes and running with their dogs.

"I'd like to take you to meet my sister, Victoria," Tony said. A small child, his pants covered with dirt, skidded his

Big Wheel right in front of them, sending up a thick dust cloud.

Erin coughed and nodded, noticing the dust didn't appear to bother her uncle in the least.

One road turned into another, winding by houses that looked more and more alike, though older and smaller than Julia and Tony's. The adobe walls were painted cream or brown, or in a few cases, a sort of salmon pink. Every yard had a beehive-shaped adobe oven that Erin knew was used to bake bread.

"This is the oldest part of the pueblo," Tony explained. "These houses were built several hundred years ago. Ours was built more recently, by the government. They decided that the concrete block was more economical."

The way he said it gave Erin the impression he thought the government didn't know what it was doing. She wondered if Julia agreed.

Tony finally stopped in front of a two-story adobe house, leaned on the wall and called out, "Hello, anybody home?"

A brown and cream-colored German shepherd ran around from the back of the house, at once the host and protector of the property.

"Hello, Tiya," Tony said, and reached out to pat him. "Paolo must be home if you're here."

The screen door opened and a young man who looked perhaps a year or two older than Erin, came out onto the porch. "Hello, Uncle."

Tony pushed open the gate and guided Erin into the yard. "Paolo, this is Julia's niece, Erin. Erin, this is my oldest nephew, Paolo."

Paolo looked at her and nodded but if he said anything, Erin didn't hear it.

"Is your mother home?" Tony asked.

Paolo nodded again. "In the kitchen," he said, his eyes never leaving Erin.

Erin followed Tony into the house. The rooms were small and rather dark, far more like what Erin had expected a pueblo home to be. A woman came in from the kitchen, greeted Tony with words she did not understand and then smiled warmly at Erin. She was shorter than Tony and very pretty, with a round face, and thick, black shoulder-length hair.

"Welcome to the pueblo, Erin," Victoria said. "Neyse has been *so* excited about your coming; she's talked of nothing else for weeks." She introduced Erin to Alonzo and Luis, Paolo's two younger brothers. A shy young girl, about six years old, stood behind Alonzo, peeking out at the visitors. Like her mother, she had long, straight hair.

"I'm Felicia," she said, her big black eyes staring unblinkingly at Erin. "Are you my cousin, too?"

Erin smiled at her while the boys laughed. "I guess I am, sort of."

"Come in, sit down," Victoria said. "Are you hungry? Can I get you something to eat?"

"No," Tony said, "we're fine. I'm taking Erin on a quick tour of the pueblo."

"Where is everyone else?"

"Julia took her class on their final field trip; Neyse and Carlos went along to help out. Fifteen seven-year-olds can be a handful, even for Julia."

"She probably won't feel much like cooking when she gets home. Why don't you come here for dinner?"

"Oh, Victoria, you don't have to do that."

"No problem. I've got a pot of chili going and there are three loaves of bread in the horno."

"Well, if you're making bread," he said with a smile, "okay."

"Wonderful."

Erin's phone pinged, telling her she had a text message. She pulled it out and read: "Give it a chance, Dorothy—it isn't Kansas after all. Taylor."

Erin grinned and sent back, "Right."

As they talked, Erin moved ever so slightly so that she could see Paolo through the small window. He sat on the adobe wall in front of the house, the dog not far from him. His dark, solemn face fascinated her and she couldn't stop looking at him. He was about the same height as Tony, a little thinner, with dark skin and midnight-black hair. She had been attracted to guys at school and, of course, had dated. But never before had she seen a guy so utterly gorgeous. She was enthralled: wondering, watching, and waiting for—she wasn't sure what.

Victoria answered the phone that Erin supposed must have rung, though she hadn't really heard it. She listened to the conversation as if in a trance.

"Well, I suppose I could come and get you," Victoria was saying.

"What's the matter?" Tony interrupted.

Erin turned to listen.

"Something's wrong with our car. Ramon took it to the garage in Albuquerque and the mechanic says he'll have to leave it overnight."

"I can pick him up," Tony offered.

"That would be great, if you don't mind. Paolo has to work at the Recreation Center this afternoon or I'd send him."

"Maybe Paolo will show Erin around," Tony said. He called his nephew into the house, while Victoria made the arrangements with Ramon. "Your father has to leave his car in town to be worked on and I'm going to go pick him up. Would you mind showing Erin around?"

Paolo hesitated. "I have to work this afternoon."

Erin immediately sensed his reluctance and felt both self-conscious and defensive. "Don't worry about me. I think I can find my way around."

"Nonsense." Victoria hung up the phone. "Paolo, you can spend an hour or so with Erin and still have plenty of time to get to work by five."

Paolo answered his mother with a shrug.

Erin saw a sullenness, an unwillingness that was more than merely not wanting to do as his mother asked. She considered telling Tony she would rather go with him, then decided against it. In the first place, it might be awkward; in the second, she was curious about Paolo. She *wanted* to go with him, even if he didn't seem delighted with the prospect.

Her phone pinged again with a response from Taylor. "Good girl ☺" She smiled, only then noting the sour look on Paolo's face.

Tony surprised Erin by giving her a quick hug. "I won't be long. Paolo will take good care of you."

Erin nodded, hardly noticing as he said goodbye to his sister and left.

Paolo looked at her with an unhidden resentment and asked, "Are you ready?"

An uneasiness came over her. "You really don't have to take me. I can look around on my own," she said, more sharply than she'd intended.

Victoria started to answer but Paolo spoke first. "It would be better if you didn't."

Erin had a feeling that he meant more than the words said, but she couldn't decipher what.

The screen door slamming behind them brought Tiya bounding to Paolo's side. He was right at Paolo's heels as the young man opened the gate, and stayed close beside him as they started out. But once the shepherd realized where they were headed, he ran off to greet three boys playing tag. Erin was sorry to see him go. It felt as if the dog were a neutral ally.

Not very far from Paolo's yard, the two young people came to a huge adobe building. It was decidedly different from any others Erin had seen; round and flat topped, and conspicuously posted NO ADMITTANCE.

"This is the *Kiva*," Paolo said. "Religious and tribal

meetings are held inside. It's a very private place."

The brusque tone of his voice stopped her from asking any questions, even though she was filled with curiosity. What did *Kiva* mean? What were the meetings about? Did he go to the meetings? Did Tony and Julia? With no doors or windows, how did anyone get inside? What was actually in the *Kiva*?

Paolo continued walking, not slowing in the slightest as they passed the *Kiva*, its walls rising at least five feet above Erin's head. She felt like a child being dragged along by her older brother.

She glanced back and saw NO ADMITTANCE again. It reminded her of the sign she'd seen earlier. She wanted to ask about it but she couldn't think of a good way to phrase the question. Finally she said, "There was a long list of regulations on a sign where we drove into the pueblo."

He glanced sideways at her, his look disdainful. "It's way too long."

"Oh? What should be omitted?"

"All of it. It ought to read PRIVATE PROPERTY: KEEP OUT."

Erin was startled by his tone of voice.

"People seem to think this is a zoo. 'Come see the Indians, see them dance, hear them sing, watch them do tricks.' It's disgusting!"

He sounded like he was accusing her. "What people?"

"Whites, mostly. Outsiders. Anybody who doesn't belong here."

His voice quite clearly said the whole world. Erin wished that she'd gone with Tony. This wasn't fun.

"Not all white people are bad," she said finally, not quite knowing what else to say.

"I didn't say they were bad, I just said they should stay away from us. There's no reason for them to be here."

Erin suddenly felt very far from her family and very alone. Her throat tightened and she fought against the tears that threatened to spill onto her cheeks.

"Don't you have to get back pretty soon?" she asked, trying to sound natural.

Her strained tone caught his attention. "Hey, look, I didn't mean you."

"I'm white."

"Yeah, well, I meant white in general."

She looked away from him toward the low foothills, wishing that she could fly there, away from him; better yet, she wished she could fly home.

"I'm sorry," he said harshly.

"You don't sound like it," she snapped back at him, and she began to run down the road.

"Wait a minute!" he called after her.

She ran, crying, not knowing where she was going, just wanting to be away from this rude and impossibly attractive stranger.

Paolo caught up with her and, grabbing her, pulled her to a stop.

"Look, I'm sorry," he said impatiently. "Really, I am," he added more softly. "I don't have any cause to be angry with you."

Tiya ran straight into Erin and nearly knocked her over. He jumped up, whining, wiggling, and begging for her attention.

Paolo looked from Erin to the dog and back at Erin, smiling slyly. "See, even Tiya is sorry."

Erin bent and patted the dog, unable to think of anything to say. She still hurt from his bitter outburst, even though she couldn't imagine it was because of anything she'd done. Tiya was delighted with the attention and wagged his tail happily as she scratched him behind the ears.

"Let me take you down by the canal," Paolo told her.

She was relieved when he broke the awkward silence, for she hadn't known how to do it herself. She nodded in agreement. "Okay."

Paolo led her between two houses and down a short alley that ended within sight of the canal. They walked across a small wooden bridge, just barely wide enough for a car to pass over. The water below was so clear that Erin could see gossamer filaments of green algae clinging to the submerged rocks. She could imagine exactly how the cold, green scum would feel under her bare feet.

"The Pueblo Indians are farmers," Paolo told her as he opened the gate that led to the fields.

"What crops do you grow here?" Erin asked, relieved now that his attitude was more pleasant.

"Our main crops are maize, that's corn, and beans. We used to grow all our food except, of course, for meat, but not as many people farm as used to. Some of them have jobs in town." He said the last as if it tasted unpleasant.

Erin didn't respond so Paolo went on.

"All the land here belongs to the pueblo. Any family that wants to farm has to request a plot from the council. If they have a field one year and don't use it or care for it, they may not get one the following year." Although most of the gardens were well tended, a few looked as if they had been abandoned for months. One sad sunflower hung its head mournfully, the shrubs surrounding it brown and neglected.

"Water has always been scarce here," Paolo said, greeting an old man who was carefully tending his crops. "We've learned to use it wisely and not waste it. There are strict rules regarding water. *All* plants are watered from the canal; if someone wants a garden by their house, the water must be carried there from the canal."

Ah, no wonder Aunt Julia and Uncle Tony don't have a lawn.

"Do you have a field?"

"Yes, right next to Antonio and Julia's." Paolo pointed toward a densely planted area. "Both my parents like farming. Besides that, with four kids, raising vegetables helps cut down on the grocery bills."

"I suppose four kids eat a lot."

Paolo looked at her curiously. "Are you an only child?"

"No, I have an older brother, Scott, but he was away at college last year. Sometimes, I feel like an only child."

"Do you miss him?"

"Yes," Erin said softly, feeling the strange, sad loneliness again.

"What's he doing this summer?"

"He's in northern California working for my mother's brother, who owns a construction company."

Chatting comfortably enough after their bad start, Erin and Paolo walked south along the canal until they came to another bridge.

"Are you allowed to swim in the canal?" Erin asked. She had noticed that in some spots the water was fairly deep and ran swiftly.

"Yes, but the current is strong in some places. It can be dangerous, especially for the younger kids."

"Do you swim here?"

The look he shot her told Erin that she'd said something she shouldn't have. "Sometimes," he said, in a tone that ended the conversation.

Her phone pinged again with another text message. This one was from her friend Heather. "U there yet?"

She texted back, "Yes. More later," and put the phone back in her pocket.

Paolo shook his head and walked away. She followed him across a bridge and west toward the mountains, wondering what she had done that had irritated him. Certainly it couldn't have been because she'd asked him about swimming in the canal. She tried to remember what they'd talked about before that. Just the fields and the crops. He continued walking in silence and she could only wonder.

The road curved slightly to the south, away from the clustered adobes toward a more open area with only a few scattered houses. Some of the yards had animal pens; the cows and goats lent a pastoral atmosphere to the pueblo.

They came to a long, rectangular, two-story adobe building, quite clearly a church, with a traditional steeple and cross overhead. Paolo looked at it with contempt. "Another outside influence we could have done without."

Erin was afraid to say anything, for fear it would set him off, but Paolo needed no encouragement.

"We have always been a religious people. Our beliefs are simple, but they're harmonious with the earth. They've survived for centuries and suit us well."

He looked at her as if expecting her to challenge him. She responded with wide, surprised eyes, merely listening.

"The Catholic Church came along and told us we'd had it all wrong—as if it were any business of theirs. Oppressor and oppressed, just like every other imperialist power."

Erin wasn't sure she understood exactly what he was saying.

"They used guilt and manipulation techniques to turn a simple people into a confused people. Children were put into parochial schools, away from their families, and then brainwashed. They were told that they must forget the old ways and accept the new way: *The* way. What right did they have?" he said, his voice rising. "We had our own..."

He looked at Erin and she watched surprise register on his face as he realized how he'd exposed himself.

"Tiya!" he suddenly yelled at the dog, who had chosen that moment to prove his dominance over a smaller animal. The shepherd growled menacingly at the other dog but obeyed Paolo, coming reluctantly to his side.

As they continued on their way, Erin thought vaguely that he and the dog were two of a kind; growling at a perceived enemy. The church was behind them now, Paolo not having said another word as they walked by. The silence felt heavy and uncomfortable. She hoped that he wouldn't flare up again. By comparison, the silence was definitely better.

She was surprised but not unhappy when she realized that she had come full circle and was, once again, at Neyse's house.

Paolo stopped short of the gate.

"The door will be unlocked," he said, without explanation.

"Thank you for showing me around."

He shrugged as if to say, *what choice did I have?*

She turned and unlatched the gate. When she looked up, he was already running down the road, the dog beside him.

"See you later," she thought she heard him say.

"I hope so," she said to herself, still not knowing quite why.

Leaning against the gate, she thought about it for a moment.

It was his eyes, she decided. *His words were angry but his eyes said something different. His eyes said there's more about him than he wants me to know.*

Those deep, dark eyes held a mystery she didn't think she could ignore.

CHAPTER 2

The house was indeed unlocked, yet another contrast with California, where people locked not only their homes, but their cars, their gas tanks, and their briefcases. Erin wondered if Indians were more trustworthy than Caucasians, a thought which, if true, unhappily reinforced Paolo's tirade against the evil White Man. Erin pushed aside the possibility as she closed the door behind her. Everyone *she* knew was honest.

She took out her phone and texted Heather, "This place is desolate. All dirt and no green. Trip may be a big mistake. Met my uncle's nephew who seems to hate whites. Welcome to the pueblo."

Heather texted back immediately. "Bummer."

Since it was clear that nobody was home, she began to look around the house, shyly peeking into Julia and Tony's room, feeling slightly guilty about her curiosity. Their four-poster bed and matching dresser reminded her of furniture she'd seen in antique stores back home when she'd gone shopping with her mother. A plain white nubby bedspread with a yellow stain on one edge hung down to the floor. The crude paintings and drawings that hung on the walls were most likely done either by Neyse or Carlos when they were younger.

Then a photograph on the dresser caught her attention and she crept into the room to get a closer look. A woman smiled at her, a light-haired beauty, probably not much older than herself. She looked familiar, although Erin had no idea who she was. Her eyes were easily her most striking feature, large, tilted upwards and piercingly clear. Erin reached out instinctively and touched the glass with her fingertips. Judging by the softened gray-brown tones of the photograph, Erin knew it was old. *I'll have to ask Aunt Julia who she is.*

She left the room and wandered into Carlos's bedroom. He hadn't made his bed that morning and the sheets looked like he'd had a restless sleep or a nightmare. Clothes and games were strewn about telling Erin he'd either left in a hurry or he wasn't as careful about his things as Neyse; the scene contrasted strikingly with her perfectly kept space.

In the entry beside the front door was a tall narrow bookcase, stacked with books of all shapes and sizes. She discovered several about the Pueblo Indians, as well as a few on Indian pottery, baskets, jewelry, and legends, and was about to reach for an oversized art book when she heard a car come into the driveway. She opened the front door and stepped out onto the porch just as Neyse jumped out of the car.

"Hi," Neyse cried, giving Erin a warm hug. "How are you? I'm so glad that you're here. Sorry I wasn't at the airport to meet you but I'd promised Mom I'd help her before we knew what day you were going to arrive." She talked in a flowing line of words, without any pauses at the end of phrases.

"That's okay."

"What have you been doing? Where's Daddy?"

Julia and Carlos came up to the porch and after mutual greetings the four crowded through the doorway into the house, talking together about Erin's arrival, the school trip,

Tony going to pick up Ramon, and Victoria's invitation to dinner. Everyone chattered at once, asking Erin questions but not giving her a chance to answer.

"Are you unpacked?" Neyse asked.

Erin shook her head. "I didn't know where to put my things."

"I emptied out three drawers for you. Come on."

Neyse led the way into her bedroom with Erin and Carlos not far behind.

Neyse's "dresser" was a set of six drawers under the bottom bunk bed that Erin hadn't even noticed. Neyse showed her the drawers she'd emptied for Erin to use.

"Did Paolo take you to the lake?" Carlos asked.

"No, we didn't have a lot of time."

"You'll like the lake. You like to swim, don't you?"

"Of course she does, silly. She's on the swim team," Neyse said impatiently.

"*Was* on the team," Erin corrected. "That shoulder injury last fall pretty well ended my career."

"Does it still bother you?" Neyse asked with concern.

"The shoulder or having to quit the team?"

"Either…both."

Erin shrugged. "Not really," she confessed. "It's healed now, and I think I was ready for a change."

Erin opened her suitcases and put her clothes into the dresser carefully, organizing them as best she could in the three drawers.

"Maybe we can go to the lake tomorrow," Carlos said. "Would you like to, Erin?"

Erin smiled at her cousin. His nose was narrower and finer than his father's and his hair was not coarse or thick like Tony's. Yet he had Tony's coloring, his black hair and eyes, and rich dark skin.

"Sure, we can go to the lake if that's what everyone else wants to do. How far away is it, anyway?"

"About five minutes. We go all the time," Carlos said.

“Well, not all the time,” Neyse corrected him. “The water’s still pretty cold but it warms up during the summer.”

Julia came into the room. “Your dad and Ramon just got back. As soon as you girls are ready we can go on over.”

“We’re ready,” Neyse told her, and then looking at Erin, said, “Right?”

Erin smiled and nodded.

Julia quickly put a variety of vegetables into an expandable net bag and they left. Carlos ran ahead of Julia and the two girls, yelling things back at them that they could scarcely understand. Erin looked at her aunt as they walked along, noting the strong resemblance to her father. She was about the same height as Erin. Sandy blonde hair curled softly around her face, setting off the wide-set green eyes that were their family’s heritage. Julia was nearly forty and while there were strands of gray in her hair, Erin saw no lines on the creamy smooth face.

Her aunt turned and smiled at her and Erin was immediately reminded of the woman in the picture frame. She started to ask about it, then changed her mind. She didn’t want Julia to think she’d been snooping around the house. One thing was certain—the woman had to be one of their relatives. She’d ask Julia about her another time.

As they approached Victoria’s house, Erin wondered if Paolo would be back from work. She hadn’t had time to ask Neyse about him. She thought about him and couldn’t repress a smile, no matter that he was rude; he was still gorgeous.

Paolo’s father, Ramon, greeted Erin with a warm smile. “I understand that Tony deserted you in order to come and pick me up. I hope that Paolo gave you a complete tour of the pueblo.”

“Yes, he did,” she said, feeling her face flush slightly. They all waited quietly for her to say something more. “I really liked it down by the canal. The fields are so green and beautiful, I could have stayed there all day.”

"That's great," Ramon said enthusiastically. "We'll put you to work in the garden tomorrow."

When they laughed, she felt comfortable and at ease among them.

Carlos went to find Luis and Alonzo, and Julia went into the kitchen to prepare the vegetables. Victoria asked Felicia to finish setting the table and when she complained, Neyse offered to help.

"*Nachra,*" Victoria said, and Erin wondered what that meant. She made a mental note to ask Neyse.

When they all sat down at the table to eat dinner, Erin was disappointed that Paolo hadn't returned. No one said anything about him and she certainly didn't want to call attention to herself by asking. As Victoria passed steaming bowls of chili, Erin discovered that either because of the hour, the walk, or the aroma, she was very hungry.

"Have some bread," Victoria urged her. She had sliced the loaf while it was still warm and put the fragrant pieces in a basket lined with a kitchen towel.

"Mmmmm...." Erin said, smelling it.

Tony smiled. "My sister makes the best bread on the pueblo," he said, giving Victoria an approving look.

No one disputed his statement.

Julia handed Erin the salad that she'd made after they arrived. It was heavy with lettuce, tomatoes, cucumbers, onion, peppers, carrots, and radishes.

"Are all these vegetables from your garden?" Erin asked.

Her aunt nodded. "Every one."

"Do you grow vegetables at your house?" Felicia asked Erin.

"No. We have a lot of flowers, but no vegetables."

"Do you eat flowers?" the child asked, a perplexed look on her face.

The family laughed.

"No, I think my mother just likes the way they look in

the yard," Erin told her.

Felicia nodded but the frown on her face indicated that she thought growing flowers was rather strange.

After dinner, Erin and Neyse helped Victoria and Julia clear the dishes.

"Shall I put the chili away, Auntie?" Neyse asked.

"No, I'll let it simmer so that it will be hot when Paolo gets home."

"What time does he get off work tonight?" Julia asked.

"He stays late on Fridays, so he won't be home until about nine-thirty."

Erin glanced at the clock on the stove. It was fifteen before eight. *I hope we stay until then.*

Luis got out a picture puzzle and set up a card table in the living room. Neyse and Erin joined the other cousins and the evening sped by so quickly that Erin was completely surprised when Paolo walked in the front door.

"Hi," Neyse greeted him. "How was work?"

"Not too many people there tonight."

He looked at Erin uncertainly, as if wondering whether his outburst that afternoon still upset her.

After a few silent moments Neyse said, "You two did meet, didn't you?"

"Yes," Paolo said. "I took her for a walk around the village. There isn't much here to impress a city girl."

"I liked it," Erin said quickly, wondering why he was starting in again. "Besides, I don't live in the city, I live in a small town."

"I thought California was one giant city."

"What do you know about California?" Neyse asked. "You've never been there."

Erin didn't say anything. She didn't want to get into an argument, so she just let Neyse do the talking, but it upset her that Paolo was being unpleasant again. What upset her most was that she didn't understand why.

"Have you eaten?" Neyse asked him.

He shook his head.

"Your mom kept the chili hot for you."

"Sounds good. I'm hungry."

He went out to the kitchen.

"*Tzanawani*," Neyse said quietly to Erin.

Erin looked at her quizzically.

"That's our word for a grump or grouch. Sometimes the Indian words are perfect for what I want to say so I use them instead of English."

"That's so neat. Will you teach me?"

"Sure. Not that we'll be able to carry on a conversation, but I can teach you a few words and phrases."

"Say it for me again."

"*Tzanawani,*" Neyse said slowly.

"*Tzanawani*," Erin repeated. "It's a good word."

Paolo came back into the room with a bowl of chili and began working on the puzzle, a well-worn picture of a mountain lake surrounded by hills of tiny purple wild flowers. She felt uneasy being near Paolo despite her attraction—or perhaps because of it. After a while, he put the bowl down and pulled Felicia onto his lap. He whispered something in her ear that made her giggle so much that he had to struggle to hold onto her while he finished eating. When he was done, Felicia put her arms around his neck and begged him to give her a ride on his shoulders.

"In a while," he promised.

Julia came into the room. "Are you kids ready to go home?"

"Not yet. Let us finish the puzzle first," Carlos said.

"Aren't you tired?" Julia asked.

"No, not really."

"Well, your father and I are going along. That trip today was exhausting."

"Can Carlos stay over?" Luis asked.

"Well..." Julia hesitated.

Having come in behind her sister-in-law, Victoria

overheard the question. “Why not? There’s nothing going on tomorrow.”

“All right. We’ll probably go to the lake, so don’t sleep in too late,” Julia cautioned.

“Let’s all go,” Neyse cried.

The cousins all agreed, with the exception of Paolo, who said nothing.

“I think that would be fun,” Julia said. “We can take a picnic lunch. From what I heard on the news, it isn’t going to be all that hot.”

“Will you go with us, too?” Felicia asked Paolo.

“I’m not sure, little one. I may have to work for Mr. Montoya.”

She put her head on his shoulder. “It’s more fun when you go, Paolo. Then I can swim.”

“You’re good enough you can swim by yourself now.”

She shook her head. “Not without you.”

Paolo hugged her and whispered something in her ear that made her smile. Erin couldn’t help wishing he’d whisper something in *her* ear.

Tony and Julia left, and Victoria and Ramon joined the children to finish the puzzle.

“You don’t have a piece in your pocket, do you, Erin?” Ramon asked when they were almost done.

“No,” she said, a little surprised.

“Bet she does,” Alonzo argued.

“I don’t!” she protested. “Look for yourself.”

Ramon grinned at Alonzo. “Check her shoes.”

Alonzo made a grab for her but she twisted away. “Okay,” she said, joining into the spirit of the game, “you’re right, I took it.” She grinned at them. “I ate it.”

Felicia’s eyes opened wide. “You ate a puzzle piece?”

Paolo chuckled, deep in his throat, and winked at Erin, amused.

Alonzo leaned over and whispered in his sister’s ear.

Her eyes widened even farther. “She does?” she said,

looking at Erin in amazement.

Neyse hit Alonzo playfully. “Don’t tell stories, ’Lon.”

“Me?” he said innocently.

“Alonzo said she eats flowers *and* puzzles,” Felicia said in a loud whisper.

“Everybody from California does,” Paolo told her.

Erin shook her head as the others laughed. She leaned down to Felicia and said, “Your brothers are only teasing you.”

“That’s right,” Neyse confirmed.

Victoria nodded to her daughter. “I just found the last piece.” She placed the final piece of blue sky into the puzzle.

Felicia looked at Paolo. “Were you teasing?”

He looked caught. “I guess we were mistaken,” he told her, giving her a squeeze.

She looked at him with complete devotion. Paolo, in turn, gave Erin an innocent smile that implied he could do no wrong. Erin had to laugh. *He certainly can be sweet when he wants to be.*

Neyse nudged Erin. “We’d better get going.”

“Paolo can walk you home,” Ramon told the girls.

“Oh, Uncle, Paolo doesn’t need to walk us home. We’re big girls. Besides, he’s afraid that the River Men might come and get him on the way back.”

Paolo put his sister down and grabbed Neyse around the waist, tickling her. “I’ll manage,” he said as she struggled to free herself.

“What about my ride?” Felicia reminded him.

He picked her up. “I’ll take you to bed, instead.”

She settled contentedly in his arms and he carried her up to her room.

A summer moon hung in the star-filled sky like a glowing oriental lantern surrounded by fireflies. The night, warm and still, was full of the sounds of the earth. Back home Erin could always hear the hum of highway traffic at night; here there was only nature’s song. So few lights were shining in curtained windows that Erin felt like she'd landed on a deserted island.

Running ahead of the three of them, Tiya began to bark, and the spell was broken.

"Tiya!" Paolo called out in a harsh, deep voice. The dog came to him immediately. Paolo said something else that Erin did not understand but which brought the animal to heel quietly beside him.

"*Tiya* means dog in our language," Neyse said.

Paolo gave his cousin a sharp look.

"Erin wants to learn a few words."

Erin sensed his unspoken disapproval and felt a sudden chill. He obviously didn't think Neyse should be teaching her their language, but she couldn't understand why. She felt as if she'd trespassed on private property, his property, and he didn't want her there.

Suddenly Erin's phone pinged with a text message. Paolo scowled at her so she ignored it. She was relieved when they reached Neyse's house and Paolo abruptly turned to go.

"See ya," he mumbled.

"*Tzanawani,*" Neyse whispered, again.

Erin nodded her agreement, liking the sound of the word.

They entered the dark, quiet house and crept into Neyse's bedroom, where someone had left on a night-light. The girls changed into their nightclothes, got themselves tall glasses of milk, and sat on the bottom bunk together talking in hushed tones.

"Well, do you like it here?" Neyse asked.

"I've only been here for one day."

"Do you like it so far, then?"

"Love it." Erin laughed.

"What did you see today?" Neyse asked.

"The *Kiva*, the canal, the fields, the church..."

"That was a good start. You'll like the lake. It'll be fun

tomorrow. Oh, and we'll have to take you to the rec center, too."

"Neyse, you said something tonight about the River Men. Who are they?"

"The River Men are, well—they're dancers, sort of. It's hard to explain. They appear in the village every year on Santa Cruz Day—that's May third. We're told they come up out of the river."

Erin frowned.

"They're dressed sort of like bums, except that they wear scary masks. They go around the village with big sacks, asking for bread. They carry whips that they crack to scare the children. If a kid is sassy or has been bad, he gets hit. Lots of little kids are really afraid of them. In fact, some of them even hide when the River Men are in the village."

"They whip little kids?" Erin asked, still frowning.

"I didn't say 'whip.' It's not all that bad. If the kid is disrespectful, he has to be taught obedience. The Pueblo has its ways...our traditions. We all have to learn."

"What else do the River Men do?"

"During the dancing, they encourage the dancers and see to it that their costumes are in good repair. Sometimes they even make fun of the dancers and imitate them. They can be funny as well as scary."

"You said that they live in the river?"

"Yeah, well, of course that's impossible but when you're little, you don't think of that. It's like Santa Claus coming down the chimney; you just accept it, that's all. Then, when you figure it out, you go along with it because of the other little kids."

"Then they wouldn't really get Paolo?"

"No, I was just giving him a bad time."

"Did they ever hit you?"

"No, but Carlos was acting fresh a couple of years ago and they gave him a swat with a whip. He hid behind Mom but they whipped him anyway. Her too, in fact."

"Her too?" Erin asked in amazement.

"Yes, and she said it hurt."

She giggled, as if the joke were on her mother.

"How many River Men are there?"

Neyse said, "Four or five." She took a long drink of the cold milk. "Say, you haven't told me about your graduation. Did you have a big party?"

"Did we ever," Erin said, and then gave Neyse the details. "A big group of us went up to San Francisco and had a dance at the Palace Hotel. It was beautiful. We stayed the night but of course, none of us got any sleep. What about you, Neys'? How was your graduation?"

"Nothing as elaborate as that. The ceremony was in the evening because of the heat, and yet two girls passed out anyway. Afterwards we had a party in the gym. It was simple but it was nice." She yawned.

"I'm tired, too," Erin said, crawling up on the top bunk.

"Sleep tight," Neyse said. "Let me know if you need anything."

"Don't worry about me. I'll be fine. G'night."

Erin lay there for a long time, unable to relax thinking of the River Men. *If they whip little children for being bad, what would they do to me for bringing a camera? Maybe I should figure a way to send it back home.*

Maybe she wasn't going to like it here after all. No doubt about it—it wasn't Kansas.

CHAPTER 3

The morning was already warm when Erin woke, her nightshirt wrapped around her body like a giant bandage, twisted so tightly she could hardly move. She straightened it slowly, trying not to shake the bed too much in case Neyse was still asleep. Then she heard her cousin stir in the bed below.

"Good morning," the dark-haired girl said.

Erin peeked over the side of the bunk. "Good morning, if it is still morning. It's so bright outside."

"It's only nine-thirty," Neyse replied, stretching.

"Nine-thirty!"

Neyse looked up, surprised. "Are you in a hurry?"

"No, it's just that, well, shouldn't we get up?"

"If you want to."

Erin sat up and stretched.

"Isn't everybody else up?"

"Mom and Dad are. I haven't heard Carlos come home yet. He'll probably stay over at Aunt Victoria's as long as he can."

They heard a soft knock at the door.

"Come in," Neyse said.

Julia opened the door and looked in. "Good morning," she said, smiling at them in a way that told Erin sleeping in until nine-thirty was perfectly all right.

"Good morning," they both said.

"Well, Erin, how did you sleep?"

"Great. I didn't wake up once."

"I'm glad. Are you hungry? How about some breakfast?"

"We'll fix something, Mom," Neyse said, pushing the sheet down with shapely brown legs.

"Okay," Julia said.

The girls both got up.

"What are you going to wear?" Erin asked.

Neyse paused for a moment. "My bathing suit and a cover-up, I guess."

Erin pulled open one of the drawers under the bed, looked through the stacks of clothes, and found her bathing suit, shorts, and a pullover. She slipped out of her shirt and into her clothes. Neyse changed into a bikini and a bright turquoise terrycloth cover-up.

After they'd dressed, Neyse stood as erect as possible next to her cousin. "I'm almost as tall as you," she said. "Tall is good," she added with a grin.

"Tall is tall," Erin stated. "Sometimes clothes that fit right are hard to find. Not to mention boys that fit right. I meant to tell you, I love your hair. It's so much longer than it was last summer. It looks wonderful on you."

Neyse nodded. "Thanks. It's grown a lot this year. I'm glad. I was tired of it short."

Neyse had Tony's dark complexion and brown eyes, but, like Carlos, her hair was fine. She was a beautiful girl, with flawless skin, an enchanting smile, and teeth just crooked enough to be believable. Her hair was straight and full, cut blunt at the ends. Although she was slightly shorter than Erin, her body was perfectly proportioned, maturely rounded, far more woman than girl.

They went into the kitchen, poured themselves juice, and fixed bowls of cereal. As they sat eating their breakfast, Carlos and Luis came bounding in. They were the same age and looked so much alike they could have been brothers.

"Well," Carlos demanded, "are you ready to go?"

"Not quite, Speedo," Neyse said.

"Why is everybody so slow this morning?"

"The lake won't dry up, Carlos."

"At the rate you're going, it will."

Erin was keenly disappointed when Paolo didn't show up with the others. *I hope he comes to the lake later*, she thought as she brushed her hair one last time.

Before long they were all packed into Tony's truck, the four adults crowded into the cab, the six cousins in the back. *This would never happen at home*, Erin mused. The trip to the lake took less than ten minutes for it was just over the hill from the pueblo.

Erin was surprised at how different the pueblo lake looked compared to mountain lakes in California. Here, the pale blue water was surrounded by sandy white beaches and light brown rolling hills. Shrubs of forest- and gray-green grew high on the hills, dense in places at the top, more scattered toward the water line. But there were no trees; the lakes in California were nearly all surrounded by trees. The area around the pueblo lake was as desert-like as the miles between it and Albuquerque. Erin thought it looked barren and desolate…almost depressing.

Tony parked in the designated area, already crowded with cars, trucks, and trailers. On the beaches to the left, several families had spread towels and blankets, and erected something for shade, including one large red and white umbrella. Tony pointed to a sandy area far to the right. "Let's go down there."

Carlos, Luis, and Alonzo ran down to the beach, threw their bundles on the sand and continued right on into the water. Tony and Ramon carried the food and a cooler. Julia,

Victoria, and the three girls took the blankets and towels. They spread everything out in a clear spot not far up from where the boys were swimming.

"Let's go into the water," Felicia coaxed. "Come on, Neyse. Come on, Erin."

They pulled off their sandals, shorts, and shirts, and followed the others into the lake where it was roped off for swimmers.

Erin made a face. "It's freezing."

"No, it's not," Neyse argued. "It just takes a while to get used to it."

They splashed and played, slowly letting their bodies adjust to the temperature. Neyse encouraged Felicia to put her face in the water, the latest thing Paolo had taught her.

"Will you race us?" Carlos asked Erin.

"She's here for a vacation," his sister said.

"That's all right," Erin told her.

"Oh, good," he yelled.

"Shall we swim out to the buoys?" Erin suggested.

"Yes! Go!" Carlos yelled, whereupon he and Luis and Alonzo began to swim furiously away from the girls. Erin took off after them, swimming easily and winning easily, yet mindful not to boast or tease him. Her father had told her before she'd left home that the Indian attitude toward competition was much different from their own.

He had explained that the Pueblo Indians had existed for centuries in a harsh, dry, unfriendly land; a land where cooperation meant survival; a land where what little water and food there was had to be shared. In the good times, everyone lived well, in the bad, all suffered equally. Over hundreds of years, the competitive urge had nearly disappeared, which was, her father said, one reason many Indians found such difficulty adjusting to white society.

"You have a good, strong stroke, Carlos."

"Yours is better. Can you show me how to swim faster?" he asked, treading water.

"I can try."

While the others swam and played, Erin coached Carlos, showing him how to improve his stroke and how to breathe more efficiently. Then, while he practiced, she swam back to the beach.

"How is the water?" Tony asked as the girls reached for their towels.

"Perfect," Neyse replied at the exact time Erin said, "Cold."

Tony laughed. "Sounds as reliable as a weather report."

Erin shrugged. "It's a little cold at first…"

"…but when you get used to it, it's perfect," Neyse finished.

The girls grinned proudly and gave each other a high five.

"A weather report or a political speech," Julia added.

Tony sighed heavily. "Guess I'll just have to go see for myself." He walked down to the water's edge, stuck his foot in, pulled it out immediately, and walked back to the blanket. "You're right, Erin. Too cold. I'll try it again next month."

"Oh, Daddy," Neyse cried, throwing herself into his lap.

"Get off me," he protested, but he held her and wouldn't let her go.

Neyse laughed and squealed, struggling, though not very hard, to free herself. "Help me, Erin."

As Erin made a halfhearted attempt to pull her cousin out of Tony's arms, her uncle suddenly let go of Neyse and grabbed Erin instead. He tickled her and teased her, pretending to let her go, then restraining her if she tried to get up.

Breathless she finally gasped, "I give, I give." and he released her so suddenly that she nearly fell over.

This was more like the uncle she remembered, and she wondered again why he'd been so reserved with her the day before.

"We'll get you later, Daddy," Neyse threatened.

"Oh?" he said, jumping to his feet.

Both girls screamed, "No," and ran, laughing, down toward the lake.

"Truce," Neyse yelled to her father.

He sat down on the blanket looking smug, and, Erin thought, very Indian.

"We give, Daddy, okay?"

He thought for a moment and then nodded.

Grinning, aware that the truce might be broken at any moment, the girls darted for the blanket that lay out of his reach, on the other side of Julia, Victoria, and Ramon.

Felicia came running up the beach and plopped down onto the blanket next to Ramon. The child was covered with mud. Erin watched Ramon, shaking his head only slightly, pick up his youngest child, who had dirtied the blanket, carry her down to the water's edge, and tenderly bathe the mud from her body. It struck her as strange that there was no scolding or harsh words to the "thoughtless child" as there might have been on a California beach.

The boys came out of the water and lay down directly on the sand. Erin's frown brought a smile to Carlos's face.

"It's warmer this way," he explained.

They lay sprawled in the sun, drinking in the desert heat. Erin took another look around the area, liking it better than she had when they'd first arrived. The water itself was beautiful, so clear that the clouds seemed to have fallen in and sunk to the bottom. She still thought it would look better with some trees, but it did have its own particular charm.

Every few minutes Erin subtly glanced up toward the parking lot, hoping to see Paolo driving in or walking down from the road. Each new group that arrived caught her attention, and each time she was disappointed. She wondered where he was and what he was doing, but didn't ask.

She listened to the others talking. Sometimes the adults would speak to the children in phrases she could not understand. They responded both in dialect and in English, one coming as naturally as the other. She liked the sound of the pueblo language spoken with a dancing of the voice. She noticed that Carlos, especially, spoke English in the same halting way, stopping many syllables completely before going on to the next. She tried to understand what they were talking about but she could not. She began to understand, however, why outsiders might want to record their speech.

Thinking about recording reminded her of her hidden camera and she wondered again what she should do about it. Perhaps she should ask Julia. Or, maybe she should wrap it up and say it was a gift she wanted to mail home.

"Erin, you look like you're in another world," Neyse said.

Erin started. "I guess I was."

"What are you thinking so hard about?"

"Oh, nothing much."

"I hope you're not homesick."

Erin laughed. "I just got here."

"I wouldn't want you to leave."

I wouldn't want to be asked to leave, she thought, the camera still very much on her mind. "I like it here, at least what I've seen."

"There's still a lot more to come. I want you to see Tent Rocks, the old village, the other pueblos that are close by, the dances, Taos, Albuquerque, Santa Fe, Dixon's Farm...."

"Wait! That's enough. I'm drowning."

"Where do you think you'd like to start?"

"I'm curious about Taos. My dad has talked about it a lot. A friend of his lives there; I think they went to college together. Anyway, Daddy says it's beautiful there."

"It really is. It's up in the mountains, far to the north. The Pueblo of Taos is interesting because the Indians there live in the exact same adobes that they've lived in for centuries. It's a popular place for tourists. Of course, you're not a tourist, you're family."

Victoria and Julia began unpacking the picnic lunch

they had prepared: fresh tortillas, green chilies, chicken and chips. They pulled every imaginable kind of soft drink out of the cooler and encouraged the girls to join them.

"It seems like we've just had breakfast," Erin said.

"Not really," Julia noted, looking at the sun. "It's about two o'clock."

"Time must pass faster here than it does at home."

"Oh, it does," Tony said. "We're always at least an hour ahead of you Californians."

She laughed, knowing that he was talking about the time difference between the coast and the southwest.

Erin kept watching for Paolo but he never appeared. He might have gone to work. On the other hand, he might not have been interested in seeing her again.

After lunch they rested and when the sun got too hot, they swam again. Erin gave Carlos a few more pointers about his style, and he practiced diligently while she floated lazily on her back, watching the clouds form into puffy white shapes like cotton batting animals.

"I'm thirsty," Neyse said, floating beside her. "Want a Coke?"

"Sure."

They returned to the beach and while Neyse looked in the cooler for soft drinks, Erin dried herself with a worn, white towel. She winced as she pulled it roughly across her shoulders.

"Erin, you're looking pretty red," Julia said.

Erin poked a finger into her leg, pulled it back quickly, and watched the pale spot turn a harsh crimson. She cringed.

"Yes, I'm getting a tan." She gave her aunt a sideways glance. "I used sun screen."

"That may be but you're getting burned. We'd better get you out of the sun. It's different here than in California. This desert sun is unforgiving."

"I'm not burned," Erin protested, knowing full well she was being unreasonable. But Julia had already begun to pack away the leftovers from lunch.

Erin smiled to herself. *Mothers are the same in New Mexico as they are in California. A little color and they think you've got a second-degree burn.*

"Besides," Tony added, "I still need to do some work in the field this afternoon."

"Can we help?" Erin asked.

"Of course; as long as we can find you a hat."

Later, after Julia had seen to it that every inch of Erin's bright pink skin was covered, Tony, Erin, and Neyse walked to the field together. Erin felt as if she were dressed for an African safari in a pair of Julia's baggy tan pants, a loose cotton blouse, and a wide-brimmed straw hat.

They carried water from the canal to the field, pulled weeds, and picked squash, tomatoes, carrots, and string beans. Warm from the sun, the fragrant aroma of the giant tomatoes made Erin's mouth water. Together they shared easy banter with other families working in their fields. When Erin was introduced to all within voice range, she felt like a sort of celebrity. Her blonde hair and fair skin were undeniable marks of an outsider that drew the attention of all who saw her. Still, they treated her graciously and she couldn't help but feel special.

She heard a dog barking and glanced up to see Tiya at the fence with Paolo. Immediately aware of how comical she must look, she wished she had argued with Julia about the clothes.

"Hi," Neyse called to him. "Finished at Mr. Montoya's?"

"Yes."

He opened the big cross-wire fence and walked through, closing it again behind him and leaving Tiya outside. Tiya whined at him but Paolo merely told him to lie down. He pulled off his shirt, tossed it over a stake at the side of the plot and began to work an adjacent field. Erin tried not to stare at him, but couldn't ignore the beautiful, muscular shoulders, probably well developed from hard work.

"You missed a lot of fun today," Neyse told him.

"Oh?"

She chattered on but he didn't respond. "You sure are quiet," she said finally.

"I'm tired. I've been helping Mr. Montoya all day."

Tony set aside a few of the vegetables and put them on the ground near Paolo's shirt. "Give those to your mother."

"*Hueh,*" Paolo told him.

"That means thank you," Neyse translated for Erin.

"*Hueh,*" Erin repeated. She said it quietly ten times. "Now it's mine forever."

"What?"

"The word—if you say a new word ten times you won't forget it. I read that somewhere and it works."

"That's a neat idea. The only thing is that if you say thank you, you'd say *nachra*. Men say *hueh*, women say *nachra.*"

"Oh. *Nachra,*" Erin repeated, trying to remember if that was one of the words Victoria had said the night before. Again, she repeated it ten times. "I'll remember them both."

"I'll give you a quiz at the end of the week and see how you do."

Erin's phone pinged and Paolo looked over, obviously annoyed. Erin sent a quick reply to Heather, and then put it on vibrate. Paolo eyed her and shook his head.

What is his problem?

Before long they were done with the necessary chores.

"Ready?" Tony asked, giving each of the girls a few vegetables to carry.

Erin looked at Neyse, wishing that her cousin would stay and talk to Paolo. She hoped her disappointment didn't show when Neyse piped, "We're ready."

They each said goodbye to Paolo, who barely looked up from his work. Once outside the fence, Erin patted Tiya, lagging behind Neyse and Tony for a few moments. As she straightened up to leave she glanced back at Paolo one last time and caught him watching her, a whisper of a smile on his face. Her stomach tumbled madly and her breath came in a rush. She wanted to smile back but his eyes went immediately to his work. However, his smile sent her home with a joyful step.

After dinner that night, as Tony and Julia sat at the table talking with Carlos and the girls, Neyse's cell phone rang. She answered it, laughed softly and, smiling, went into her room.

"That's Juan," Carlos told Erin. "Neyse's in love with him."

"Don't talk behind her back," Julia said in a firm, yet kind, voice.

"Well, he's her boyfriend."

"Yes, he is."

"And she talks to him all the time."

"Frequently," Julia agreed. "How is your sunburn, Erin?"

"Oh, it's not bad," she said, though it was beginning to sting.

"I have some aloe you can put on after you bathe if you'd like. It helps me. I burn fairly easily too, especially the first time I go out each summer."

Julia's skin was pale and creamy and Erin had trouble imagining her with a tan.

"Thanks. I'd like to keep from peeling if I can."

"It's a good thing you covered up before working in the field," Tony observed, "or you'd look like a cherry drop tonight."

Erin chuckled.

Neyse came back into the living room saying, "That was Juan."

"As if we couldn't have guessed," her brother said.

"He's coming over, Mom, okay?"

"Of course."

Less than five minutes later, Erin heard a distinct, loud bird trilling outside. Neyse jumped up to answer the door and let Juan in. He seemed quite tall, perhaps because he was so thin. He had glossy black hair and smooth dark skin, just like Paolo, but any similarity ended there. He was relaxed and gregarious, joking with Neyse and the family.

When Neyse introduced him to Erin he smiled broadly. "Thank goodness you're finally here," he greeted her. "I was afraid that while Neyse was waiting she would have a heart attack from prolonged anticipation."

He shared an easy conversation with both Tony and Julia, not in the least strained or contrived, and asked Erin several questions about California, particularly about her hometown. His speech was less lilting than anyone else's, more like what she was accustomed to at home. She wondered why. *He sounds like a politician*, she finally decided with an inward smile.

"Well, what do you think of the pueblo?" he asked, "not that you've had enough time to decide or, for that matter, would tell us if your impressions were bad."

"I like what I've seen, but you're right, I haven't been here long enough to look around very much."

"She's learning the language," Neyse bragged.

"Oh really?" Juan said. *"Hatza shra ketra?"*

"What?" Erin said, frowning.

Although everyone laughed, Erin did not have the feeling they were making fun of her, merely teasing.

"I'm still on lesson one. Give me a week or so and I'll be able to answer you."

Erin liked Juan. In a way, he reminded her of her brother Scott. When she told him so, Neyse agreed.

"He must be a wonderful person," Juan said with an impish grin.

Erin nodded but Neyse shook her head. "Actually, he's much nicer than you," she told him.

"Who wants dessert?" Julia asked.

"Everybody wants dessert," Tony said. He looked at the others. "Why wouldn't we want dessert?"

Julia dished up ice cream and brought the bowls, along with a jar of thick chocolate syrup, to the table.

"What do you have planned for the week?" Juan asked, more than halfway finished with his sundae.

"The usual Red Carpet tour," Julia said.

Carlos began to laugh. "Yeah, we'll have Erin sit on an old rug and we'll drag her around everywhere." He nearly fell off his chair laughing at his own joke.

Neyse shook her head in dismay but began laughing herself, watching her brother's glee. "He gets like this," she told Erin.

Carlos nearly knocked his empty plate on the floor.

"*Heem'e,*" Julia cautioned.

"That means enough," Neyse translated.

"Let me know if you need a tour guide," Juan offered, adding, "I've lived near here all my life." He began to laugh as well.

Julia and Tony looked at each other and at the same time said, "It's time for bed."

The young people collapsed in laughter. Tony and Julia shook their heads. Tony put his arm around his wife, they both said goodnight and went into their room.

"Anybody want to watch TV?" Carlos asked.

"Sure," Juan said.

Carlos tuned in an action show and they all watched, paying little attention to it. Juan kept them distracted with running comments about the program, particularly the commercials. Carlos finally went to bed before it was even over, leaving the other three laughing, and not knowing what was going on.

"I suppose I'd better go," Juan said a short time later. He looked at Erin. "Glad to meet you finally. I'm sure I'll see you again soon."

Erin grinned at him. "Nice to meet you, too."

While Neyse walked outside with him to say goodbye, Erin took a quick shower, applied the aloe gel her aunt had left out for her, put on her nightshirt, and climbed into bed. When Neyse came back in, Erin asked, "Have you been seeing him for long?"

"Only for about six months. I didn't really notice him all that much until he started at U.N.M. He changed then. He's more interesting to me now."

"U.N.M.?"

"The University of New Mexico…in Albuquerque."

"Is that where you'll go?"

"Yes. After I finish at the junior college."

"Does Paolo go there?"

"No. Since they graduated he's been doing odd jobs around here along with working at the rec. He's *so* stubborn; he thinks that going to college is selling out to the whites."

"Wow. He really doesn't like Caucasians."

"Yeah. And, it's sad because he won't be satisfied for long doing odd jobs. Juan keeps trying to convince him to enroll, but he hasn't made any progress yet."

Erin noticed her sweet smile when she spoke of Juan. "Do you like him a lot?"

"Yes, a whole lot."

Erin nodded. "I can see why. He's nice."

"Would you mind if he comes along with us when we go places? Not everywhere, of course, but sometimes?"

"Of course not."

"Oh, good. You'll like him, Erin. He really is fun."

Erin laughed. "I noticed."

I just hope he likes me. I'd hate to face another one like Paolo.

CHAPTER 4

The radio was on, spewing forth music that just didn't seem to fit on the pueblo—a loud, hard rock that shook the very air in the room. Neyse was dancing around the kitchen, keeping time to the beat while she poured juice and made toast. Erin liked some rock, but she didn't much like the song. As a rule, she preferred something a little mellower first thing in the morning.

"I wish you'd turn that down a little," Julia called from the living room, her voice edged with irritation.

Neyse rolled her eyes and made a face at Erin, but turned the volume down as her mother had asked.

"*Nachra*," Julia said, sounding relieved.

Tony came through the kitchen, grabbed a heavy lunch pail off the counter, and gave each of the girls a pat on the head. "Have fun today."

"Be careful," Julia cautioned him as he kissed her.

He worked on a highway construction crew and had been called unexpectedly to work on his day off. He was rushing because the crew couldn't start until he arrived.

Julia grimaced when the truck raced out of the driveway, but said nothing.

It had been a week since Erin's arrival, a week that had gone by quickly, nearly every minute packed with something new and exciting. When her parents had called the night before, Erin found it difficult to relate the details of all the things she'd seen: the lake, the canal, the fields, the hills, the delicious food, the family, the friends, and the fast-changing weather.

In a way, she'd wanted to tell them about Paolo. But everyone stood around her as she talked and she didn't know what to say about him. Even if she'd been alone, she decided she probably still wouldn't have known, for although she'd seen him twice since the afternoon they'd worked at the field, she was confused and uncertain about him.

"Juan will be here before long," Neyse said, pulling Erin's thoughts back to the kitchen.

"I suppose I'd better get dressed, then."

"Jeans would be best. It'll be hot but you won't have to worry about getting scratched."

Erin thought a minute then chuckled. "Better cooked than hooked?"

Neyse gave her a blank stare.

"You know, caught on a thorn. Hooked?"

Neyse shook her head.

Erin shrugged. "It's still early. What can I say?"

Neyse, Erin, and Juan planned to hike to Tent Rocks, a geological formation located not far from the village.

Tony had driven Erin to see them early one evening during the previous week and she'd been fascinated. When Juan heard about her enthusiasm, he'd suggested a hike.

Erin changed into jeans and a bright blue and white plaid cotton blouse. She braided her hair, tying the braids with strips of fawn-colored leather Carlos had given her. Neyse came into the bedroom just as Erin finished dressing.

"Gee, you look cute."

"Yeah, particularly my nose," Erin said as she rubbed it unhappily.

"Well, it shouldn't peel again. But, you'd better put a lot of sunscreen on before we leave."

"Juan's here," Carlos yelled from the living room.

Neyse shook her head, grinning. "He's a regular butler."

Julia tapped on the door, opening it slightly. "I assume you heard your brother."

Neyse grinned and nodded.

"Auntie, do you have any sunscreen?" Erin asked. "I'm afraid if my nose burns any more, it'll peel right off my face."

Julia nodded. "I've got just the thing."

Erin followed her into the bathroom, where Julia applied what Erin felt sure was the most obnoxious smelling ointment imaginable, to what remained of her nose.

"I know, it stinks, but it doesn't hurt, right?"

Erin nodded her head, agreeing unhappily.

"And the sun absolutely can't get through it."

"Nothing in its right mind would come close enough to get through it."

"It'll be better in a couple of days."

"Erin," Neyse called, "are you ready?"

"Yes," she said, her voice directed toward the living room. "Thanks, Auntie, I think."

The three young people bid goodbye to Julia and Carlos and headed west, toward the hills, which were the opposite direction from the canal.

"Is that beauty cream on your nose, Erin?" Juan teased her.

She hit him on the arm. "You're mean."

He sniffed loudly. "It has a distinct aroma."

Erin leaned close to him. "If you like it so much, let me give you a kiss," she suggested, only a little surprised at her own boldness. She liked Juan, just as Neyse had said she would, and she already felt comfortable teasing him.

"That's okay," he said, pushing her away. "Maybe another time."

Juan carried the backpack with their lunch and a towel in it. Both Neyse and Erin had offered to carry it, but he'd put it on without comment. Erin detected a stubbornness in him similar to that which Paolo displayed openly.

"It's too bad that Paolo couldn't make it today," Juan said, almost as if he'd heard her thinking about him. "I was hoping that he could trade his hours at the center with someone else, but he couldn't find anyone who'd help him out."

The girls looked at each other. "Paolo?" Neyse said, surprised. "Who asked him?"

"I did. I didn't think you'd mind."

Neyse had told Erin that Paolo and Juan were not only the same age, but best friends. She guessed it was natural for him to want to include Paolo.

Neyse shrugged. "No."

"He said maybe he could make it next time," Juan added.

Erin hadn't told Neyse that she was interested in Paolo and she couldn't imagine that Juan knew it, either. She considered what Paolo had said and wondered whether he had really meant it. It sounded like one of those answers people give when they don't know what else to say. Still, Paolo didn't strike her as the type who would say something simply for the sake of being polite. He'd certainly made it a point to be impolite on more than one occasion.

She thought about the previous afternoon, when Julia had realized that there weren't enough vegetables for dinner and Erin had offered to run to the field to get some squash. She had indeed run, looking back toward the hills instead of watching the road, and she had literally run right into Paolo.

"Is that the way you jog in California?" he'd asked, clearly aggravated.

His unexpected attack had startled her. "I'm sorry," she'd mumbled.

"It's no wonder you lead the nation in freeway accidents."

She'd frowned at him. "I said I was sorry."

"Don't worry about it, I'll live."

Her phone beeped but she ignored it. He gave her a disgusted look.

She shook her head. "What is it with you? Why are you always so grouchy?"

"Guess I'm just on the warpath," he'd said sourly, and walked away.

"I guess you are," she had called after him, adding, "*Tzanawani,*" delighted she'd remembered the word. His step had faltered just enough to let her know he'd heard her, but he'd kept on going.

She wondered how he would have acted if he'd come with them to Tent Rocks. Surely he wouldn't have been nasty to her in front of Juan and Neyse. She was disappointed, thinking that she'd lost a chance to be with him under pleasant conditions for a change.

The road was dry and dusty, bordered on either side with prickly, gray shrubs. The bushes looked lifeless and dreary, as if all their energy had been expended in surviving, with nothing left for beauty; determined, ornery bushes, hardened to life, sort of like Paolo.

Juan pointed at the foothills to the west. "See the piñon trees?"

Erin squinted and looked. "Where?"

"Right there," he emphasized.

"I don't see any trees," she insisted.

"Come on, Erin, they're right on the side of the hills."

"You mean those little bushes?"

"They're not bushes, they're trees," Juan said again.

Erin grinned impishly at Neyse. "Piñon bushes."

Neyse laughed. "Well, I'll admit they aren't exactly giant redwoods, but we still call them trees."

Neyse proceeded to tell Juan about the trip she and Erin's family had taken the previous summer to the redwood forest, where they had seen trees wider than a car and hundreds of feet high.

"Must be real impressive," he agreed. "I'm afraid that I can't offer you anything quite that dramatic, at least around here." He thought for a moment. "The mesas west of Albuquerque are breathtaking. Maybe we could take you there."

"I wasn't complaining," Erin said, feeling guilty that he'd taken her so seriously.

"No," he said, holding up his hand. "Don't apologize. I want to give you something you can write about in English class in the fall. 'What I Did Last Summer,' by Erin Fraser."

Both Erin and Neyse laughed.

He continued to point out other plants as they walked along, telling Erin their Indian names and their uses.

"How long have your people lived here?" Erin asked.

"About seven hundred years. Before that they lived in the old village to the north."

Erin frowned slightly. "Why did they move from that village to this one?"

Juan looked thoughtful. "Nobody really knows for sure. There aren't any written records. The old village is on the top of a high mesa, a flat mountain. It could be they decided they didn't need such an impenetrable spot. Or, it could be that the path of the river changed and they needed to be closer to the water. It's hard to say. Lots of theories, but that's all they are."

Erin liked the way Juan answered her questions. He made her feel as if he wanted her to know instead of reacting the way Paolo did. She couldn't understand Paolo's attitude, but she couldn't help hoping she could change it.

The warm day was heavy with the scent of sage that seemed to fill Erin's lungs before she could draw in enough air. She panted in short, sharp breaths, slowly letting her body get used to the pungent fragrance. She remembered the smell from summer camp when she was twelve, a memory that made her smile. She'd fought going away alone that summer, nearly cried when she'd left on the bus. But she'd loved camp and had truly cried when the bus pulled away from the mountain lodge. Although the desert surroundings here were nothing like the mountain camp, the sharp smell was exactly the same.

They came to a narrow valley carved in between the foothills. A riverbed, almost dry, ran down the center of the valley. The growth was heavier near the water and considerably greener than it was on the hills.

Ahead and to the right Erin saw the area called Tent Rocks, hills cut away and formed into what looked like inverted ice cream cones or teepees.

"We're here," Neyse announced.

Layers of rock, sand and earth, plainly visible, looked as if they had been painted on the hills in stripes of brown, gray, and gold. There were many of the strangely shaped forms, some perhaps thirty feet tall, almost every one with a large boulder on the top.

Erin shook her head in wonder. "I've never seen anything like it."

They continued walking until they came to a road that cut off to the right toward the Tent Rocks. Juan led the way up the road; Neyse and Erin followed not far behind.

"You doing okay?" Juan asked.

"Yes," they replied, sounding not too convincing.

The sun beat down steadily and the gray sandy earth glared back. Erin was hot and thirsty and hoped they would soon stop, but Juan kept right on walking. Erin had never liked to hike, a fact established at camp. "Just one more step," a counselor had told her those many years before, though she could easily see that one more step wouldn't get her anywhere. She resented the obvious ploy. She loved seeing new places but felt, somehow, that she was being punished if she had to walk there.

She looked again at the oddly-shaped hills. "How did they get like that?" she asked.

"I don't know that, either," Juan said sounding apologetic. "It was a slow process—most likely over thousands of years. This is a river valley. The water very possibly brought the rocks down from the north and over time, the water and wind together eroded the hills. The upper areas would have been affected the most and that would explain the tent-like shapes. The rocks at the top probably acted as a sort of defense for that 'tent'."

They climbed to the foot of the hills, where Juan slipped the straps of the pack down off his shoulders and guided the pack onto the ground in front of him.

"I'm tired," he announced.

"What a relief!" Neyse said with a sigh.

"I thought you'd never stop," Erin said.

"Why didn't you say something?" he asked, looking first at Neyse and then at Erin. "This isn't a marathon, after all."

"Fine," Neyse said. "Then let's stay here the rest of the day."

"But we've got to go up to the end of the canyon," Juan argued.

The two girls looked at each other and started laughing.

"Let us rest a little while, okay?" Neyse asked.

"Sure," Juan agreed. "Want something to drink?"

When they nodded, he took water bottles out of the backpack, opened them and offered one to each of the girls.

"*Nachra*," Erin said, pleased that she had again remembered the right word.

Neyse nudged Juan in the ribs and grinned at him, delighted that all her teaching was paying off.

Erin took a long swallow and sighed with pleasure.

"Amazing how good water can taste, isn't it?" Juan said.

Both girls agreed as they sat down to rest and enjoy the drink.

"Our word for water is *tzitzi*," Juan told Erin.

"*Tzitzi*," she repeated. "Good word to know on a hot afternoon."

Juan glanced around. "I'm going to look for some piñon nuts."

"What are they?" Erin asked.

"They're seeds from the piñon trees," Neyse told her. "Our people have eaten them for centuries. They're about the size of your little fingernail and very sweet. You'd like them, I think."

The girls relaxed while Juan looked around on the ground under the piñon trees. He returned with a half of a dozen cones and showed Erin how to shake them to get out the nuts. He offered some to Erin, who put one in her mouth.

"Oh, come on, Erin. Eat some," he encouraged her.

She took several more and ate them, nodding her head. "They're good."

They were soft, tasted sweeter than sunflower seeds and more delicate than peanuts.

Looking pleased at his success, Juan went to search for more.

"Do you come up here very often?" Erin asked Neyse.

"You mean Juan and me?"

"Yes."

"No, not really. We brought a picnic up once during the spring, with Paolo and Alonzo and a friend of his, but other than that, no. My mom likes to come up here, especially in the evening."

Erin remembered how pretty it had been the evening Tony had driven them here, the sun swimming in a pool of pink and orange foam. As it set, the hills turned from gray to purple to blue, like rotating filters on a camera. Erin thought that evening light was much softer and prettier than daylight, and kinder, here, to the harsh desert landscape.

Juan appeared suddenly and dropped a pile of cones in front of the girls. "Snack time," he announced and began to shake out the nuts.

It seemed to Erin that it took forever to extract enough nuts to make a handful. She wondered how practical they could be as a food for a large family. She smiled, inwardly, imagining a pueblo woman spending two weeks in the hills gathering enough nuts for the family to snack on during their favorite TV show.

"Ready to travel?" Juan asked.

Neyse looked at Erin. "What do you think?"

"Okay," she agreed, secretly hoping that they wouldn't go too much farther before lunch. The nuts had made her stomach growl and she was beginning to feel very hungry.

As they continued up the canyon, they heard the sounds of the stream, the insects, and the wind in the trees. The day was beautiful, clear and hot, a breeze blowing just enough to cool their faces. Juan began tossing piñon nuts into the air and catching them in his mouth. Neyse tried it as well and was fairly successful. Erin discovered she lost more on the ground than she found in her mouth, so she gave up.

"I wonder who first learned that these nuts were edible," she mused.

"Our people found long ago how to survive in these surroundings...how to take the gifts of the earth and use them. We've taken our food and our cures from the earth for centuries. For years our traditions were considered quaint rituals by outsiders, but now science is learning that there are good reasons for our practices." Juan looked pleased but not smug. "Our religious ceremonies are sacred and, therefore, secret, but much of our other knowledge could be used by outsiders if they would be willing to learn from it."

"Do you have cures for diseases?"

"Nothing yet for the common cold," Juan said with a grin, "but we do have herbs and roots that we use for healing."

"Is there a medicine man in the pueblo?"

"We don't talk about that."

He said it matter-of-factly and without emotion but Erin felt chastened. She was on forbidden ground again and she didn't like the feeling.

Juan nudged her. "It's okay. You didn't say anything wrong. There are just certain subjects we don't talk about."

"I'm sorry."

"Don't be. Outsiders have always been curious about our ways. To us, certain things are sacred and very private. Ask me whatever you want. You'll just have to understand that there are some things I won't talk about."

"It isn't just outsiders," Neyse added. "There are things that are just for the men and they won't even discuss them with pueblo women."

"Really?" Erin frowned. "Aren't you curious?"

"We're taught not to be," Neyse said flatly.

Erin couldn't imagine not being curious, couldn't imagine not wondering about such secrets. "It doesn't seem fair to exclude the women."

"Why not?" Juan challenged.

"Because women have the same rights as men to...to...whatever they want."

"In your world, maybe," Juan said. "Not in ours."

Erin looked at Neyse. Her cousin gave her a glance that said, *Don't push it.*

Just then a huge black cloud passed overhead, turning everything dark.

Erin looked up. "Do you think it will rain?" she asked, happy to change the subject.

Juan shrugged. "Maybe. As hot as it is, it would be a pleasure."

But after a time, the cloud moved on and the sun blazed again.

"The weather changes so quickly here that we often don't know what it will do from one minute to the next,"

Neyse told her.

Juan found a shady spot in a clump of piñon trees, spread a frayed blanket on the ground and indicated for the girls to sit. He pulled the lunch out of the backpack—tuna sandwiches on dark rye bread that Neyse and Julia had made. Although the drinks were no longer very cold, they were nonetheless wet and refreshing. They ate greedily, drank deeply, and talked very little.

"This sure tastes good," Juan said, halfway through his sandwich.

Erin agreed.

"A natural talent," Neyse said with a sassy smile.

"Well, since you did such a good job on the lunch, I volunteer to do the dishes," Juan told her.

"Aren't you wonderful?" Neyse returned with a loving sigh.

Erin took off her shoes and socks and ran her feet deep into the sand to cool them off. "Mmmmm, feels good."

"Don't get too relaxed," Juan told her. "We've got a lot more hiking to do."

"Oh, Juan," Neyse moaned, "I think that the trip back home is about as much as I can handle."

"Well, let's at least get a closer look at the formations."

The girls agreed, if somewhat reluctantly.

"After we rest a while," Neyse said laying back, her arms behind her head.

Both Erin and Neyse catnapped while Juan took a quick walk down to the riverbank. By the time he got back, the girls had cleaned up the remains of the lunch. Juan led them along the foot of the hills, following the jagged, turning path back toward the pueblo.

"You two look like real adventurers," he said. "I'd love to have a picture. Did you bring a camera, Neyse?"

"No."

"You, Erin?"

Erin tripped and nearly fell. "No," she said nervously.

Neyse glanced over at her. "What's the matter?"

"Nothing," she said, hoping they would change the subject.

They walked for a long time in silence. Erin avoided looking at them.

"Are you sure you're all right?" Neyse asked again.

"I'm fine."

"You sure don't sound fine."

"Well, I am." She was quiet for a moment.

Neyse didn't let up. "Come on Erin, what's the matter?"

"Well, actually..."

"What?" Neyse pushed.

"I did bring a camera."

"Where is it?" Neyse asked, looking utterly confused.

Erin's stomach turned and her mouth went dry. Certainly if the pueblo women didn't have many rights, an outsider would have even fewer.

Neyse and Juan continued to look at her, awaiting an answer.

"It's hidden in the closet in your room," she said to Neyse.

"What?"

"I didn't know I wasn't supposed to have a camera until I saw the big sign at the entrance and then it was too late and I didn't know what to do with it," she said in a rush. "After you told me about the River Men I was afraid to say anything 'cause I was afraid that I would get into a lot of trouble or that they'd send me home."

Despite her fear, she felt relieved that it was finally out in the open. At least now the matter would be settled even if they did send her home.

Neyse and Juan looked at each other, bewildered. Finally Neyse said, "For heaven's sake, Erin, nobody's going to send you home. The sign is intended to keep outsiders from imposing on our people. Some pueblos are overrun with tourists who harass the Indians constantly." She shook her head. "You can take pictures of us. We're your family. And if you want some of the scenery, it's not a problem. You can't take any of the *Kiva* or the dances—private things. And I wouldn't walk around the pueblo alone with your camera since the people don't know you yet. But other than that, there shouldn't be any problem."

"As far as the River Men are concerned," Juan added, "they're only around on May third. Even if you showed up at the dance with a camera, *they* probably wouldn't say anything to you. More than likely one of the elders would tell you that the camera is forbidden, not the River Men. They aren't concerned with outsiders; they're concerned with our people. They aren't a police force. They're our teachers."

Relief spread over her like a cool breeze. "Thank heavens; I was so scared."

"You looked scared," Juan agreed.

"Erin, please," Neyse said, "if you're worried about something or don't understand our ways, ask me, okay?"

"Okay."

As they continued home, Erin tried to figure out how to ask Neyse what she wondered about the most—Paolo.

CHAPTER 5

"Victoria's on the phone," Neyse called to Erin. "She's going up to Taos to visit Ramon's aunt, who is sick, and she wants to know if we'd like to go with her."

Erin came into the living room from the bedroom where she'd been looking at Neyse's yearbook.

"What do you think?" she asked Neyse quietly.

"I think it would be fun," Neyse whispered back.

Erin nodded. "Okay, then, let's go."

"What time are you leaving?" Neyse asked her aunt. Then, after a pause, she continued. "Okay. No, that's fine. We'll get ready and come over right away."

Julia and Tony had gone to Albuquerque to take Carlos to an orthodontist appointment, leaving the girls on their own for the day.

"What are you going to wear?" Erin asked as Neyse hung up.

"Shorts. You can be sure it'll be hot."

Erin rejected three pairs of shorts before choosing her white ones. After putting them on she complained, "Seems like I wear these every day."

"I'd wear them if they looked that good on me," Neyse

complimented her.

Erin put on an Ed Hardy shirt to wear with the shorts. Neyse put on a pair of denim shorts and a white tee shirt that said, “SKI TAOS” in bright red letters.

“We really look like a couple of tourists,” Neyse said with a broad grin.

Erin shook her head. “I’d rather die.”

“Don’t worry about it. We’ll only be seeing other tourists anyway.”

Neyse texted her parents to tell them where she and Erin were going, just in case Tony and Julia returned before the girls.

“Ready?” she asked Erin.

“Ready.”

As they hurried down the road to Victoria’s, Erin felt the muscles in her chest tighten, wondering whether Paolo would be going with them. She wanted to ask Neyse more about him: whether he was ever nasty to her; whether he had a girlfriend; and what had made him dislike whites so much. But she couldn’t. Not yet.

Paolo was in the front yard chopping wood when they arrived, Tiya sitting not far away.

“Hello,” Neyse called out.

“Hi,” he said. He rested the ax for a moment and gave Erin a sort of lopsided smile, which made her stomach flip-flop. “*Dz’i qwii gudeii kwih?*”

Neyse leaned toward Erin and said, “That means ‘what’s up?’”

“We’re going to go up to Taos with your mom,” Neyse told him.

“So I heard. Touristas.”

“There are worse things.”

“Not many,” he said, returning to the woodpile.

Erin’s phone pinged with a text message from Taylor. “What’s on for today?”

She quickly texted back, “Taos Pueblo. Details later.”

She glanced up at Paolo, who was scowling at her.

"Do you sleep with that thing?"

She shook her head no, not knowing what to say. She looked at Neyse, who merely shrugged her shoulders and walked up to the screened door.

"We're here, Auntie," she called out.

"Would you come in and give me a hand for a minute?" Victoria replied from inside the house.

"Sure," Neyse said, entering the door without hesitating.

Should she follow Neyse into the house or stay outside, Erin wasn't sure.

"Where's your camera?" he asked her with a sideways glance.

"My camera?"

"Nobody goes to Taos without a camera."

She leaned against the fence. "I am."

"The villagers there have learned to perform well for the Anglos," he said, biting the last word harshly. "Well, you can always take pictures with your phone." He emphasized the last word.

She grimaced. "I'm not going for a performance," she said, her anger rising.

"Why are you going?"

"Because I was invited!" she snapped.

"Do you go everywhere you're invited?"

"If the person who invites me is pleasant. And you?"

"I only go where I want to go."

"I'll bet. And no doubt you walk over anyone in your way to get there."

As soon as the words were out of her mouth she was sorry. She knew from the look on his face that they had hurt, yet, as her mother would say, she was losing the battle.

"History shows that it's the Anglos, not the Indians, who walk over people."

"I don't think the Anglos are trying to walk over the

Indians. Not any more, anyway. We need to work together. We're all people—we're all the same."

"No. We certainly are not all the same. We're all different. Those who believe we're all the same are kidding themselves. Those who try to ignore the differences are doing a disservice to everyone. It's wrong to mix our cultures; it weakens both and strengthens neither."

You're wrong, she thought, but she didn't say it.

"Hard to dispute, isn't it?"

"People are people. We all have flesh and blood and bones. We all need to sleep and eat and work. Just because I live in a town and you live on a reservation doesn't make us different."

"A *reservation?*" He glared at her. "This is a pueblo, not a reservation!"

"Well," she stammered, realizing that she'd set him off, "it's the same thing, isn't it?"

"It certainly is *not!*"

"What's the difference?"

He looked at her with loathing. "Don't they teach you *anything* in California?"

"They teach us lots of things."

"You should have paid more attention in History. A reservation," he repeated, shaking his head. "I don't believe it."

He turned his back and began chopping wood again.

Just then, Victoria and Neyse came out of the house.

"Ready to go?" Victoria asked.

Erin nodded with relief and walked away from Paolo, knowing his eyes followed her. *Why does he have to pick a fight with me every single time we're together? What is it with him?*

Since the back seat of Victoria's ancient SUV was littered with books, toys, and bags, the girls squeezed into the front. The pale blue vehicle was so encrusted with dirt that Erin doubted it could ever be clean again. She tried to imagine what her grandfather would say about such a dirty car. He was almost fanatical about washing his sedan every weekend, insisting that people judged you by the condition of your automobile—not the model and year, but how clean it was. If that were true, those people would have a hard time living on the pueblo.

"Can I turn the radio on?" Neyse asked Victoria.

"Of course."

Since she was sitting between her aunt and her cousin, Neyse could reach the controls of the radio easily. She tuned in a station that played popular music.

The beautiful scenery along the highway could not distract Erin from what Paolo had said. She wondered if he had been talking about Julia when he mentioned mixing their races. *I've never heard him be rude to Aunt Julia. In fact, he seems to like her. I wonder if he thinks it was wrong for her to marry Tony or if he was talking about the two cultures in general.*

She tried to decide what the difference between a reservation and a pueblo might be. She could remember nothing about it from history class. Indians lived in both places; that was all she knew. Since Paolo had said that she could have learned it in school, it obviously wasn't one of the pueblo secrets. That helped. Neyse could tell her, or Juan, of course. Probably the safest bet would be Julia.

I'll ask Julia. I wonder who Julia asks when she wants to know something.

"We almost had Paolo's company today," Victoria said as she slowly guided the car past a group of boys walking along the road.

Erin turned to look at Victoria.

"Really?" Neyse said.

"I don't know what made him change his mind. What I mean is, I asked him last night if he wanted to go along today and he said no. Then this morning, when I mentioned you girls were joining me, he said he wanted to go after all."

"Well, what happened?" Neyse asked.

"He was changing his clothes when a call came from the rec center. One of the other lifeguards is sick and they really needed Paolo to work this afternoon. He told them he didn't want to, but they'd already called everyone else who had the day off. He didn't feel as if he could refuse."

"Darn," Neyse said. "That would have been fun."

Now Erin was even more confused. If Paolo wanted to go with them, why had he been so rude to her? If he disliked Erin so much, why had he decided to go along after he heard she was going? Juan had said that he wanted to go with them on the next hike to Tent Rocks. He only seemed to be interested in her when she *wasn't* around. It made no sense. In frustration, Erin pushed all thoughts of Paolo aside, thoroughly confused and unable to make any logic of it at all.

I'm never going to be able to figure him out. He's impossible.

As they drove along the highway, Erin saw the sign for the city limits of Santa Fe.

"Have you ever been to the capitol?" Victoria asked Erin.

"No. I've never been to New Mexico before this summer, and so far, I haven't been anywhere but the pueblo."

Victoria looked at Neyse. "I think she should at least see the plaza, don't you?"

Neyse agreed with an emphatic nod.

Victoria took a downtown exit off the freeway and followed the signs that led to the plaza.

"We don't have time to get out and walk around but you'll get a chance to look, anyway."

The streets were crowded, the traffic moved slowly. Erin rolled down the window hoping for a breeze, but the air hung hot and dry outside, without even a hint of a current. She wished that one of them had taken the back seat because even with the air conditioner running, her body was so warm that her legs stuck together.

Victoria drove around the plaza, a small rectangular park bordered by one-way streets.

"The Mexicans traditionally built their towns around a central plaza. It was the center not only for the government, but for the trades and social affairs as well. Many of the cities and towns in the southwest have preserved their original plazas. In a few, here and Taos for instance, they are still the center of town."

"I like the Old Town Plaza in Albuquerque the best," Neyse told Erin. "It's a beautiful park, the grass is always green, there are lots of trees, and the people are friendlier there."

"What's that?" Erin asked. The building she pointed to was long and low, and stretched for almost the entire block on the east side of the plaza.

"That's the Palace of Governors," Neyse said. "The governors actually used to live there. Now it's a museum."

All along the covered sidewalk in front of the Palace, Indians sat on small rugs, with jewelry, pottery, and moccasins lavishly displayed in front of them. Curious tourists crowded around, examining their craftsmanship and making purchases. Erin looked at the dark, serious faces as Victoria waited for the ambling pedestrians. The Indians appeared completely uninterested in prospective buyers, and spoke only when making a sale.

Erin wanted to know if anyone from Neyse's Pueblo might be here but she was reluctant to ask. Paolo had made her feel that questions were risky.

Victoria wound through the narrow streets of Santa Fe and headed back to the freeway.

"It looks like all the buildings are adobe," Erin commented as she strained to see up and down all the streets that they passed.

"The City Council would be happy to hear that,"

Victoria laughed. "Santa Fe tries very hard to preserve the look of the southwest, in commercial buildings as well as in their homes."

On the freeway once again, Erin watched the landscape with growing interest. Hills of desert sagebrush surrounded the squat adobe houses along the way. There were no lawns, no flower gardens, and no manicured yards. The style complemented the work of nature, as if it belonged that way. Erin was beginning to like it.

The country between Santa Fe and Taos was open and barren, dotted with small, gray-green piñon trees. The Rio Grande River flowed down the valley to the west of the highway, its banks lined with cottonwood trees just like those by the river at the pueblo.

"There's Camel Rock," Neyse cried. "Look, Erin, quick!" she cried, pointing out the window past Victoria.

"My gosh, it does look like a camel," Erin exclaimed as she caught sight of the large, distinctively shaped rock formation.

"When my dad was little they used to be able to climb up on it, but it's fenced off now. You can only look."

Victoria sighed. "It's very sad."

"I suppose people were painting it or something dumb like that," Erin said with disgust.

Neyse nodded.

"That happens a lot back home. I can't understand what prompts people to be so destructive. A few jerks ruin it for everyone."

"A lot of the well known monuments are fenced off now, just as a precaution. Juan went to Window Rock, in Arizona, with his cousin last spring. He said that it's fenced off, too."

"Window Rock?"

"That's the headquarters of the Navajo Nation. Juan's mother is Navajo. I've never been to Window Rock, but I've seen pictures of it. The 'window' is twice as high as a man, almost round, and created entirely by nature. We can look for a postcard in Taos so you can see it. They always take pictures of it against the sky. It's really beautiful."

They continued along the highway. Victoria pointed out the ruins of an ancient pueblo, just barely visible from the road.

"There are several deserted pueblo villages throughout the state," she told Erin. "One of the things that brings so many tourists to Taos is that the Indians there inhabit the exact same adobes where they've lived for centuries."

Erin felt as if her ears were stuffed with cotton. She yawned to make them pop, knowing that they must be gaining altitude even though the landscape was almost flat.

Looking out Victoria's window, Erin saw that they were, indeed, high in the hills. The Rio Grande Gorge dropped away dramatically in jagged cliffs and colorful outcroppings to the west of the highway.

"It almost looks like the Grand Canyon," Erin said.

Victoria smiled. "It's smaller, but yes, it does."

"Have you ever hiked there?" Erin asked Neyse.

Her cousin shook her head. "Too much work for me."

Erin nodded. "Probably for me, too."

"If you'd like, I can drop you two off at the plaza while I go to visit Tia Louisa," Victoria suggested. "Then we can meet for lunch. How does that sound?"

"Fine," Neyse agreed, then paused, noticing Erin's frown. "...doesn't it?"

"Oh, yes, that's fine."

"What's the matter?" Victoria asked Erin.

"Nothing. I was just wondering, isn't *tia* the Spanish word for aunt?"

"Yes, that's right."

"Well, why..."

"Why do Indians use Spanish words?"

Erin nodded, hoping that she hadn't asked another question that would get her into trouble.

"The Spanish were the first Europeans to come to the pueblos, actually long before the whites. Our ancestors learned the Spanish language along with the Catholic religion. The young people don't learn much Spanish any more, but the older generation still speaks it frequently. Some of us mix English, Spanish and our dialect until it is unrecognizable. The sad thing is that for years, our own native tongue wasn't taught to the children."

"Is it taught now?"

"Some schools have started teaching it. Traditionally, it's left up to the parents, mainly the mother. If she doesn't care or isn't willing to make the effort, it doesn't happen. It's an oral language and not written, so it needs to be communicated from one person directly to the next. It's sad, but some people don't think it's important, which is a shame, because if we don't preserve our heritage, it will be lost."

She sounded like Paolo. *I wonder if that's where he got his ideas. She certainly isn't nasty about it the way he is.*

Buildings along the side of the road increased in number and size as they neared the town. Victoria turned from the highway onto a street that led to the plaza, smaller than the one in Santa Fe, and to Erin, more welcoming.

Victoria pulled to the side of the street, and the girls jumped out.

"See you in a bit," she told them.

They stood momentarily on the corner, each girl taking a careful look at the small town square. Taos plaza was not as big nor as crowded as the one they'd just seen. People ambled slowly along the sidewalks. It reminded Erin of the colorful stories she'd read about Old Mexico.

"Where to first?" Neyse asked.

Erin opened her eyes wide. "I don't know. What do you think?"

"Let's start here, go around the whole plaza and look in every store."

"And not miss a thing," Erin said with a chuckle.

They went from door to door, looking in every store. They both preferred the small specialty shops and galleries that were packed with Indian and cowboy art, jewelry, pottery, and clothing, some very fine and very expensive.

They spent a long time in one gallery, both girls attracted to the artwork.

"I like that," Erin said. She pointed to a colorful print of an Indian woman of rather large proportions, prominently displayed on the central wall.

Neyse nodded. "That's by R.C. Gorman. Mom likes him, too. He's pretty famous. I've heard that some people call him the Picasso of American Indian art. He died just a few years ago."

"I really do like it," Erin said again, leaning close to it to check the price on the small, round tag. She rolled her eyes. "Yikes! It's expensive."

"You should see the price of the originals."

Erin shook her head in disappointment.

"If you're really interested, I'm sure you can get notecards printed with his drawings."

"I can?"

"I think so. I know I've seen them before. If we can't find any here, I'm sure we can get them in Albuquerque."

"Oh, let's look."

They continued around the square, skipping some stores that did not appeal to them as much as others. Finally, in a small gift shop off the main street they found note cards with Gorman prints just as Neyse had promised. Erin took her time looking through the boxes, finally narrowing her choices to just two.

"I'm going to get one box for me and one for my mom."

"Does she like stationery, too?"

"She likes everything. It's a cinch to buy things for her. My dad is something else. I never know what to get for him."

"What about a book?" Neyse suggested as Erin paid for

the cards. "He's always sending them to my mom."

"That's a good idea. He does love books. But he has so many; I wouldn't have the vaguest idea what to get him."

"We'll figure out something."

Although they looked in every shop they passed on their way, by the time they met Victoria, Erin hadn't found anything she thought her father would like. Neyse told her aunt the problem.

"You have plenty of time," Victoria encouraged Erin. "After all, you aren't leaving until the end of August, right?"

"Not until the last Sunday," Erin said.

"Well, don't worry. I'm sure you'll find the right thing before then. You haven't even been to Albuquerque yet. Also, I imagine that Julia could give you some ideas."

Neyse agreed. "Mom's good at that sort of thing."

After they'd finished lunch, Victoria paid the bill and led the way to the car.

"Do you still want to see the pueblo?" she asked Erin.

"Oh, yes. That is, if you don't mind."

"Not at all," Victoria assured her. "I think it would be interesting for you to see how different it is there. Tourists can walk around freely and take pictures as they please. It's probably one of the most photographed places in the state."

Erin thought of Paolo's remark and it stung her again. She was glad she hadn't known she could take pictures before they'd left home. She probably would have wanted to bring her camera. She did have her phone, of course, but she wasn't sure she actually wanted any pictures.

They reached the pueblo entrance in less than ten minutes. Victoria paid the fee that allowed them to drive the car inside and park. The village was much smaller than Neyse's, the adobes built much closer together, some three stories high. It looked almost like an apartment complex. Simple ladders made of wood leaned against the walls of the tallest adobes and although Erin assumed they were used for entrance, she didn't see anyone go up or down.

Tourists were everywhere, like ants crawling in and out of the doorways, swarming here and there, nearly every one with a camera. For their part, the Indians went about their business, cooking, baking bread, selling jewelry and pottery, some almost catering to the outsiders while others ignored them completely. The tourists followed the Indians around, asking questions, taking pictures, buying goods, and trying to be part of the activity. To Erin, it seemed like a sideshow, not a sharing.

As she watched it happening all around her, she felt very much an outsider, who didn't belong there. She followed Victoria into two of the low-ceilinged adobes, realizing she was not in the least interested in the jewelry the artists were selling, but unsure how to tell Neyse's aunt.

She imagined people parked in front of her home in California, trampling across the lawn, and waiting on the doorstep for someone to come out so they could take a picture.

How can they stand it, she wondered of the dark, quiet people. *The tourists are like invaders: they have no right to be here.*

I have no right to be here.

The thought startled her.

"I'd like to look inside the church," Victoria said, pulling Erin out of her reverie.

"Uh— I think I'll just sit by the river for a while," Erin said.

Neyse and Victoria headed toward the churchyard while Erin walked slowly along the bank of the river. She was glad to be alone. She found a spot on the wooden plank bridge that crossed the river at the south end of the Pueblo. The crystal clear water sparkled and sang, rushing over the smooth stones in its bed.

As she sat, her legs hanging down over the side of the bridge, a black and cream-colored puppy came wriggling up to her, whining and wagging his tail. She patted his head and he licked her hand. A small child walked down to the bank, filled a bucket with water and looked up at Erin.

He spoke to the puppy while looking at Erin, clearly saying, "He's mine."

The puppy stumbled clumsily down to the water and followed him away.

A couple with three noisy children walked across the bridge, talking loudly and yelling at the child, trying to get him to turn around.

Erin cringed, wishing that she were somewhere else.

She turned away from the unsavory scene and looked upstream, toward the distant mountains that rose high into the deep blue sky. Puffs of cotton-white clouds played tag above her head. She could see a photograph everywhere she looked.

But as beautiful as it was, it seemed to her that it was being spoiled.

At last Victoria and Neyse came out of the church. They walked over to the bridge and sat down beside Erin.

Victoria sighed, looking at the far-off mountains. "I hate to leave, but I have to get home so I can bake pies for dinner."

Erin jumped up, startling the older woman. "I'm ready."

"You are?" Victoria asked, surprised.

"Well," Erin hesitated, "I mean as soon as you are."

Victoria stood slowly, as if the bridge itself were urging her to stay. She led the way back to the car. Erin forced herself not to run; she could hardly wait to leave the place.

The girls sang with the radio, all the way home. Erin's mood had lightened considerably by the time Victoria pulled up in front of Neyse's house.

"Thank you for taking us with you," Neyse said to Victoria as she and Erin got out of the car.

Erin nodded in agreement.

"I was happy to have the company," Victoria replied.

She waved as she drove off.

Julia was preparing dinner when Neyse and Erin walked in the front door.

"You're just in time," Julia said with a delighted smile. "Will you set the table?"

"Can we have a cold drink first?" Neyse asked, a look of mock desperation on her face.

"Yes, of course," her mother said. "Erin, what did you think of Taos?"

Erin looked up from the drawer where she was counting out the silverware. "It was beautiful," she said quickly, hoping that the subject wouldn't dominate the evening's conversation.

"I'm glad. Not everybody likes it there since it's so old. What are you two up to tomorrow?"

Neyse put the plates on the table and shrugged. "We haven't even talked about it."

"Let's go up to the lake," Erin suggested, hoping to have a chance to digest her thoughts.

"Sounds good to me," Neyse said.

"Rest while you can," her mother told them, "because we have to start getting ready for Feast Day before long."

"For what?" Erin asked. They'd finished setting the table and were walking into Neyse's room.

"Feast Day," Neyse repeated. "We have it every summer. Actually, different Pueblos have it at different times of the year; ours is in July."

"But what is it?"

"It's a celebration—a dance—an Indian dance, to ask for rain for the corn. It's a custom our people have followed for hundreds of years. And, of course, a feast. We have lots of food. We ask friends and family to come and watch the dance and eat traditional food." She giggled. "We're very good at eating."

Erin hesitated and then asked. "Can I come?"

"Of course. Everybody comes."

"Outsiders?" Erin asked.

"Sure. Friends from town, and out of town as well. And tourists, too. They come here from all over. It's an open dance. Anybody can come."

"Will you dance?"

"Yes. Daddy and Carlos and I will all dance."

"What about your mom?"

"Mom can't dance."

"You mean she doesn't know how?"

"No. She's not Indian."

Neyse said it as a matter of fact, but it hit Erin in the face.

Like me. We don't quite belong.

CHAPTER 6

Erin knew before she ever opened her eyes the next morning that the day was destined to be hot. She'd learned to read the signs. Her body felt slightly heavy and almost unclean, even though she'd taken a bath the night before. She jumped out of bed.

Neyse was already up and in the shower. Erin heard her singing and grinned, glad that they were both "morning people."

A note on the dining room table told the girls that Julia had gone to town to market but there was no sign of Carlos.

"We're going to the lake today, right?" Erin asked as she finished her second bowl of fruit.

Neyse sat on the couch reading the comic strip. "Sure are. You ready?"

"I will be as soon as I put my suit on."

Neyse didn't look up.

"Well?"

"I'm coming," she said absently, still engrossed in the paper.

Erin waited but Neyse didn't move. "Let me help you," Erin said. She grabbed the paper out of Neyse's hand and ran into the bedroom.

"Better bring it back," Neyse called, "because I'm not moving until I finish it."

"You'd better come get ready or I'm going to tear the stupid thing into pieces." She glanced around quickly and found a piece of paper in the trash.

"Come on, Erin."

"Goodbye, comic strip," Erin said and swiftly ripped the scrap paper in two.

"Errrrrin!" Neyse wailed as she came dashing into the bedroom. She saw the two pieces of paper in Erin's hands and grinned. "You are a real brat."

"No, I'm not. A brat would have ripped up the comics."

"May I please finish reading in peace?"

"Of course. Be my guest. Take your time. Don't feel badly that I'm here for such a short time. By all means, read the comics."

Neyse shook her head. "You ought to go into the theater. You're a great actress. I don't even *want* to read them anymore."

"Don't let me influence you," Erin said with an innocent look.

Neyse rolled her eyes. "You're unbelievable."

They changed clothes in record time and easily hiked the distance to the lake. As they walked down the gravel boat ramp toward the beach, they saw Paolo sitting near the edge of the water, Tiya beside him, still but alert.

Neyse waved to him. "Want some company?"

He turned just enough to see them and shrugged.

Neyse shook her head. "Whatever," she mumbled.

"What about right here?" she suggested to Erin, indicating a large sandy area far enough away from Paolo that their conversation wouldn't be overheard.

"That's fine," Erin said quietly, hoping their presence wouldn't set him off. She didn't want to take any chances of causing another scene like the one the day before.

"Don't worry about him. It isn't his beach. Sometimes he's moody. It's best to just leave him alone."

Erin spread her towel and dropped her backpack on the corner to keep it from blowing over. Uneasily she pulled her phone out and put in on vibrate. As she dropped it back in her backpack she noticed Paolo shaking his head. *He really has a thing about the cell phone. But why? Everyone here has one.*

Neyse put her towel on the sand next to Erin's, pantomiming an elaborate greeting to Paolo behind his back. Erin did her best to suppress her giggles as she sat down.

"Don't you want to swim?" Neyse asked.

"No, not yet. I think I'll read for a while."

She stretched out on the towel. Pulling her book from the backpack, she opened it to the marked page and stared at the words, reading a single line five times without understanding it. Her eyes drifted across the sand to where Paolo sat. His bare shoulders rose and fell evenly with his breath, the skin dark and smooth.

Handsome doesn't really describe him, she thought. *At least not the way guys are handsome back home.* Paolo had a mysterious quality that Erin couldn't put her finger on, couldn't quite figure out. He'd been nasty to her more than once. *Ah, but when he smiles—I just can't get enough of it! And he's so sweet with Felicia.* He had a gentle side to him—one that was loving and sweet. *That's the Paolo I want to get to know.*

She imagined walking over to him, sitting down and saying, "Hi, can we be friends?" What would he say? What would he do? Her eyes darted across the sand and she realized he was watching her.

"Good book?" Neyse asked. She grinned impishly.

"Pretty good," she said, dropping her eyes.

"Let's go in the water."

"Okay."

They stood, pulled off their shorts and shirts, and ran to

the wet, water-lapped sand, each daring the other to brave the first cold plunge.

Neyse suddenly splashed in, ducked down and came up squealing. "It's freezing."

Erin waded in just past her ankles. "Let's not go into the water," she said, shivering.

Neyse scooped up an armful of water and sent it flying toward Erin, who could not move fast enough to avoid getting drenched.

She glowered at Neyse. "You didn't really mean to do that, did you?" she asked, walking slowly toward her cousin.

"As a matter of fact…" Neyse began.

"Did you?" Erin asked again, drawing closer and looking more menacing.

"No, no, I didn't," Neyse cried. "Don't, Erin, don't!"

Erin began to laugh. "Don't what?" she asked innocently.

"Don't get too close or I'll have to do it again," Neyse shrieked, flinging another wave of water at Erin.

Neyse dove into the water and began to swim away from the beach. Erin leaped after her, swimming with a strong, well-developed stroke. Within only a few body lengths, she caught Neyse's foot.

"I didn't mean it! I didn't mean it!" Neyse cried out, kicking free.

Erin felt for the bottom, but they were in water just over their heads.

They trod water, giggling at each other.

"It doesn't seem cold anymore," Erin observed.

"Especially right near the surface," Neyse agreed, floating on her back.

Erin did the same thing using a simple hand movement to help maintain her buoyancy.

She knew that Paolo was watching her. She'd seen him a couple of times when she'd glanced back at the beach, but more than that, she could feel him, his eyes boring into her.

She didn't have a sense of anger or danger, merely interest: intense interest. She shot a quick look at the beach, just to prove to herself that she was right. She was. His eyes were fixed on her and did not waver. The grin on his face said he knew she saw him.

She wanted to yell at him to quit being such a dope, but she didn't. Instead she said to Neyse, "Let's swim out to the buoys."

"I'll swim out," Neyse agreed, "and then you can pull me back."

"We'll discuss that part when we get there."

She made a silent deal with herself not to look back at Paolo until they'd reached the buoys. She imagined turning to find him swimming out after her. She imagined him close enough to touch, maybe even to kiss.

She swam slowly, setting a pace she thought Neyse could match without straining. She judged that the buoys were about one hundred meters from the beach, a distance she knew she could cover fairly quickly. Swimming slowly, for Erin, was more difficult than racing, since it went against her instincts and her training.

Neyse stopped suddenly. "Are you sure you want to do this?"

"You can't stop in the middle of an event!" Erin told her, trying to sound like her own coach.

"Why not?"

"Because the judges will disqualify you."

Neyse thought a minute. "I don't want to be a swimmer," she said. "I've decided I want to be a...a...diver."

"Impossible."

"Why?"

"There's no diving board."

Neyse nodded. "Oh," she said, and began swimming again.

Erin kept the deal with herself and didn't look for Paolo. She picked up the pace just a little as they neared the buoys, anxious to see what he might be doing. When they reached the rope, she tagged the white float and turned immediately toward the beach.

Paolo and Tiya were gone.

She searched the entire shoreline and then the hills surrounding the lake. She looked around in the water hoping Paolo might be swimming out to join her. But, they were nowhere to be seen. They'd simply vanished.

"Race you back," she said, hoping Neyse hadn't noticed her disappointment. "You start," she said, giving Neyse an advantage. But she still beat her cousin by several strokes.

Erin tossed Neyse her towel as she dragged herself out of the water.

The dark girl dried herself and then gave Erin a confident smile. "Bet I can beat you through lunch," she said breathlessly.

Erin shook her head. "No contest. I never do two events back to back."

They flopped down on the beach and dumped their lunch onto the towel in front of them.

"Is this all we brought?" Neyse asked, looking at an unadorned cheese sandwich as if it might be poisonous.

"*You* packed it," Erin reminded her.

Neyse pursed her lips. "Don't ever let me do it again."

Erin laughed at her cousin's serious tone.

They ate the sandwiches, tortilla chips and small, purple plums, and drank soft drinks that had warmed in the midday sun.

"Siesta time," Neyse said when they'd finished.

"A nap? Are you kidding?"

"Not at all." Neyse settled herself on her towel.

Erin lay back on her warm towel and looked up at the sky, squinting against the sun's glare. White clouds fringed the distant hills like a frilly petticoat showing below a dirndl skirt. As she watched, the clouds blew eastward, puffing out, expanding into lifelike forms of people and animals.

"I love to watch clouds," she said. "It's as if they're alive, the way they move, like giant animals walking across the sky. They're so beautiful."

"Beauty above me, with it I wander," Neyse mused.

"What?"

"It's part of a Navajo prayer that's in one of Mom's favorite books. As a matter of fact, I think your dad gave it to her. Anyway, it says,

'Beauty above me, with it I wander;
Beauty below me, with it I wander;
Beauty before me, with it I wander;
Beauty around me, with it I wander'."

Neyse was quiet for a moment and then explained. "Indians have always seen the connections between the earth, themselves, and the beauty of creation. Our religion reflects those connections."

"Where I live there are so many tall buildings and billboards, you can hardly see the sky, much less appreciate its beauty."

"Oh, come on, Erin..."

"Well, practically. Something new is always being built. Here it's so quiet, and peaceful, and beautiful. I think I'd like to live here."

"You're a dreamer. There's just as much to complain about here as there is in California. We have ugliness as well as beauty. You can't tell me that the ocean or the redwood forest isn't beautiful. I loved everything I saw last summer."

She smiled at her cousin. "You're right. But still, you can't deny that the clouds are beautiful, the lake is beautiful and the day is beautiful. I love it here and I'm glad that I got to come and visit."

"So am I," her cousin said with a smile.

The clouds turned dark and the wind blew cold across them. Erin pulled her shirt on over her head. She unbraided her hair and ran her fingers through the damp, golden waves.

"Do you want to go?" Neyse asked.

"Do you think it's going to rain?"

"I doubt it. It'll probably clear up in five minutes."

"Then let's stay."

"Okay," Neyse murmured, her head buried in the crook of her arm.

Erin tried to relax. Would Paolo come back? *Probably not. He was so angry yesterday. He lives in a different world. He wants it that way and I should just ignore him.* She rolled over onto her stomach and opened her book once again. Before long, she was engrossed in the story and was startled when she felt the first raindrops on her bare legs. Neyse was asleep.

"Neyse, wake up," she said, shaking her gently.

"Huh?"

"Come on. We'd better go."

Neyse jumped up and quickly stuffed her towel into the backpack. "We should hurry. Looks like it's going to pour."

They ran up the hill to the road but the rain ran ahead of them. Water dripped down their faces and soaked their hair. They looked at each other and began to laugh.

"It isn't going to do us any good to run now," Neyse laughed.

"Guess not."

They slowed down and walked with the rain, evenly and steadily. The earth soaked up the moisture like milk poured over bread. They passed the sign at the pueblo entrance and Erin smiled at the memory of her first fears. She'd been accepted by everyone she'd met and felt part of the pueblo. Then Paolo's harsh words sprang into her mind. *We are different. It is wrong to mix our cultures. It weakens each, and strengthens neither.*

She looked at her cousin. Neyse was lovely—her fine, black hair and dark, round face combined the best features of both parents. A quick smile showed beautiful, milk-white teeth. Her brown eyes sparkled. White mother, Indian father...beautiful daughter. What was so bad about that?

It's wrong, she imagined she could hear Paolo say.

She thought about Taos and about how the outsiders were spoiling its peace. *Maybe he's right. Maybe we should be kept separate.*

Neyse interrupted Erin's confused thoughts. "Looks like Mom just got back." They saw Julia carrying groceries into the house from the car.

They ran down the soggy driveway, mud clinging to their feet. They each grabbed a bag and went around to the back of the house to the kitchen door.

"Here, Mom," Neyse called out as she pulled open the door.

Julia looked up, surprised.

"You take these and we'll get the rest."

They unloaded the car and then rinsed their feet under the spigot by the back door before going into the house.

"You girls are soaked," Julia said with dismay.

"Oh, Mom," Neyse said with equal exasperation.

"You get right out of those wet clothes," Julia said, "and then you can help me put this food away."

They grabbed the shorts and shirts they'd changed out of earlier, took towels from the linen closet and closed themselves in the bathroom.

"Let's take a fast shower," Neyse said, pulling off her dripping wet shirt and suit.

They took turns, standing under the fine spray of warm water, rinsing their hair. Neyse's chocolate brown skin was creamy smooth. Erin felt jealous of the rich, dark color. She had tanned considerably since she'd arrived, and the golden brown color contrasted emphatically with the white skin unexposed to the sun. But she would never have Neyse's rich, dark color.

She put her arm next to Neyse's. "Another six months and I might be able to pass for your sister instead of your cousin."

Neyse laughed. "Not too many of us are blonde."

"No, I guess not."

She continued to think about the differences while they put away groceries, and couldn't quit even after they'd gone back into Neyse's room. They sat on the floor playing cards. The rain had stopped.

"I wonder what Paolo's problem was today?" Neyse said.

Erin felt uncomfortable, not wanting to bring up the argument she'd had with him the day before.

"Does Luis or Alonzo ever act like that?" Erin asked.

"Luis doesn't know what a bad mood is. He's always happy. Alonzo? He gets huffy sometimes, but he gets over it right away."

Neyse thought for a while. "What did you and Paolo talk about while I was helping Victoria yesterday?"

Erin hesitated. "Taos," she said finally.

Neyse looked closely at her cousin. "And he bugged you about being a tourist, right?"

Erin nodded, looking down at the cards.

"Did you get into an argument?"

Erin nodded again. "I didn't want to, but he started in on this whole bit of whites walking over the Indians and I got mad."

"Sounds like one of his Indians' rights speeches he used to give at school. Sometimes he sounds like every Anglo ever born is out to get him. Actually, Pueblo Indians had many advantages compared to lots of others. At least we weren't moved off our land."

"What do you mean?"

"Well, the Plains Indians were nomadic tribes. They had always roamed the land, following the buffalo, moving their villages when the food in the area was gone. When the government took over, the Plains Indians were confined, put on reservations to live."

Erin took a chance. "Isn't this a reservation?"

"Oh, no," Neyse said. "The Pueblo Indians already

lived in established communities. They were farmers. Our lands were well defined and no threat to the government, so we were left pretty much alone."

Erin nodded, understanding now why Paolo had made such a big deal about it when she had assumed that a pueblo and a reservation were the same thing.

"The Plains Indians were told they had to learn to farm," Neyse continued. "They were forced to give up the ways of their ancestors and live the white man's way. It must have been like going to prison."

"No wonder the Indians resent the whites."

Neyse shrugged. "Some do, some don't."

"Neys, do you feel like you are more Indian than Anglo?"

Neyse thought for a long time. "I am Indian. I don't much think about being Anglo. It would have been hard if Mom had tried to make us half and half. She's lived pretty much like other Indian women. Oh, she complains sometimes because women aren't treated all that well, but she's done it just the same."

"It seems like she's had to make a lot more changes than your dad."

"Overall, you're right. But this is the lifestyle she values, so she got something important in the bargain. She constantly tells me is that marriage is a big adjustment. She says no matter whom I marry, it will take a lot of work. She wants me to go to college before I get married. Most of the pueblo girls don't."

"Wouldn't it be fun if we could go to college together; maybe even be roommates?"

Neyse's face lit up with delight. "'Neyse and Daisy Go to College,'" she chuckled. "We can write the sequel to 'Neyse and Daisy Take a Vacation.'"

Erin rolled her eyes at the nickname.

"Would you come here or shall I go to California?" Neyse asked.

Erin thought for a minute. "Let's both go back east—Harvard or Yale."

"Perfect. Shall we tell them we're coming or just surprise them?"

They collapsed in laughter.

They spent the rest of the afternoon day dreaming about their lives in the Ivy League.

"Dinner's ready," Julia called.

As the family finished at the table, Tony announced, "I'm going to take everyone who wants to go for an ice cream for dessert."

Erin, Neyse, and Carlos were delighted. Julia decided to stay at home. All three youngsters jumped into the back of the pick-up and sat against the cab. Erin's hair whipped in the wind as they drove up the highway to the small market. When they arrived, they saw Victoria just getting out of her car in the parking lot.

"*Kuwe tzi*," Neyse called.

"Hello to you," Victoria said, smiling. "What are you up to?"

"Daddy brought us for an ice cream," Carlos told her.

"Ummmm, that sounds good."

Tony gave his sister a warm smile. "Of course, if we got a better offer…"

"Well, I did make *pakowtza* this afternoon."

"Vanilla ice cream and *pakowtza* sounds even better. Of course, we'd bring the ice cream."

Erin's heartbeat quickened. Maybe Paolo would be there. She felt a quivering excitement in her stomach.

She caught a glimpse of herself in the window of the market and the sight made her miserable. Her hair was a wild mess. She hadn't brought her purse along, much less a brush. The trip to Victoria's in the back of the truck wouldn't help matters any.

Victoria came out of the market carrying two half

gallons of milk.

"May we ride with you, Auntie?" Neyse asked, almost as if she'd read Erin's mind.

"Yes, of course," Victoria said.

Once in the car, Erin ran her fingers through her hair, trying to get out the worst of the tangles.

"You look fine," Neyse told her.

"What is it that you made today?" Erin asked Victoria, quickly changing the subject.

"*Pakowtza.* Cookies. The kids love them. Antonio's the biggest kid of all."

Neyse agreed. "Daddy eats the most, but that's because they're so good. You'll like them, Erin."

Sitting stiffly on the couch at Victoria's, she felt as nervous as if she were a thirteen-year-old at her first dance. She was glad that they had at least progressed past the "are you having a good time?" stage and that the conversation no longer focused on her.

Tony and Ramon were arguing with Victoria about a legislative candidate, a man they'd all known since their teens. The boys sat at the table working together on a model windmill Luis was building. Paolo's greeting had been cool, and Erin wished that she'd never agreed to come.

"Mama, can I have more ice cream?" Felicia asked.

"Yes, dear. Neyse, would you mind helping her?"

"Sure," Neyse said, getting up. "Come on, Erin."

Erin couldn't get into the kitchen without brushing past Paolo. He made no effort to move his chair, so she touched his shoulder as she went by.

"Look out, Paolo," Felicia said with a smile.

Paolo grabbed her and wouldn't let her go. She giggled and squirmed but he held her tightly.

"Let go, Paolo, I want my ice cream."

"*I* want your ice cream," he teased.

Erin warmed to his gentleness. While Neyse dished up

Felicia's dessert, Erin rinsed the other bowls.

"I'll take that," Paolo said to Neyse.

"*Dzah*, Paolo! It's for me," Felicia cried.

"Paolo," Victoria said. "*Ma'a*."

Paolo and Felicia struggled back and forth. Paolo suddenly loosened his grip, Felicia slipped, and the ice cream crashed to the floor.

Paolo groaned. "I'll clean it up."

He went into the kitchen to get a cloth from the sink. Erin had no way to avoid him. Standing next to him she realized how big he was, his shoulders even broader than she had thought. He smelled fresh—like soap.

Paolo looked at Erin, saying nothing, but smiling at her as if they shared a guarded secret. She shivered at the intimacy of his gaze.

Suddenly, Felicia rushed into Paolo, catching him unprepared. He slammed into Erin and she fell heavily against the wall.

Paolo jumped to her side, pulled her up, and asked, "Are you all right?" His eyes were filled with concern.

"I'm fine," she assured him, taken aback by his unexpected tenderness toward her.

"What about my ice cream?" Felicia demanded.

"In a minute," Paolo told her.

"I'll get it for you, Felicia," Neyse said, opening the freezer again.

Erin grinned at Paolo as he cleaned up the mess on the floor.

"What did Victoria say to Paolo?" she asked Neyse in a whisper.

"*Ma'a*? That means stop," her cousin whispered back. "And *dzah* means no," she added.

Erin nodded, making a mental note of the new words.

"Look, Erin," Luis called to her.

She went toward the table and again brushed past Paolo. He smiled at her again, and her stomach tumbled.

Erin had trouble concentrating on what Luis was saying, her mind still captivated by Paolo's smile.

Neyse tapped her arm. "Are you ready to go? Don't forget that the movie starts at nine o'clock."

"Okay," Erin replied reluctantly, trying not to sound disappointed.

Neyse kissed her aunt and uncle. "See you later, Daddy," she told Tony.

Paolo opened the front door. "I'll walk with you," he said.

"You don't have to," Neyse told him.

"I know," he said, calling Tiya.

The night was warm and clear, the stars so bright that they seemed close enough to touch. There were no signs of the storm earlier that day.

Paolo walked between the two girls, though slightly closer to Erin, his shoulder touching her off and on as they went along.

"Paolo," Neyse said, "remember the night we played hide and seek and Luis fell asleep down by the church?"

Paolo laughed, deep in his chest.

Neyse skipped around in front of Erin and Paolo and walked backwards, telling Erin the details. "It took us almost two hours to find him," she said, "and then he cried because everybody else had given up and gone home."

Erin giggled with her cousin.

"Oh, Paolo," Neyse cried, "let's get a gang together and play one night this week."

Erin expected Paolo to respond negatively but instead he nodded and said, "You know, I think that'd be fun. What do you say, Erin? Is the city girl too old to play hide and seek with her country cousins?"

"Just as long as I don't have to be IT first. After all, other than the church, I don't know any hiding places."

Paolo grinned at her.

"What about Saturday night?" Neyse suggested.

"That would be good for me. I don't have to work on Sunday, so I can stay out all night if I want. Don't let your folks hear about it."

"I won't," Neyse promised. "We'll tell everybody to meet by the church at eight."

They were home then and Paolo turned to go. "*Haakoh kaama,*" he said to Neyse, who replied, "*Ha'a.*"

"You're not afraid of snakes, are you, Erin?" he called over his shoulder.

"Snakes?" Erin gasped.

"Oh, don't listen to him," Neyse said.

Paolo laughed, called Tiya, and off they went.

"What did you just say?" she asked Neyse.

"He asked if he could leave. It's a way of saying goodbye. I said *ha'a,* which means yes. More words for your vocabulary list."

They went into the silent house.

"Why don't you want your parents to know about Saturday night?" Erin asked quietly.

"It's a sort of unwritten law that we can't play hide and seek at night. They think there are too many things that could happen to us. We do it. We just don't let them know about it."

Erin lay in bed after the lights were out that night thinking about Paolo and the gentle way he played with Felicia. She fell asleep imagining his strong dark arms wrapped around her.

CHAPTER 7

The sun had just dropped behind the mountains leaving ribbons of cotton-candy pink along the horizon—sky-blue-pink, as Erin remembered her grandmother saying. She and Neyse were sitting on the tailgate of the pickup truck, talking as they waited for Juan.

"I feel as if I gained ten pounds at one meal," Erin told Neyse, looking at the bulge of her usually flat stomach.

They'd had Indian tacos for dinner, a traditional pueblo dish that was growing to be one of Erin's favorites. The girls had helped Julia make the dough for the taco shells that afternoon. Indian Fry Bread was used instead of the corn tortillas Erin was accustomed to back home. It took Erin several attempts to get the knack of throwing the dough lightly on her fingers, like Julia did, to get the right size, shape, and thickness. Two tore apart and one fell on the floor. Julia had only laughed and told her to keep trying.

"I'm going to have to quit eating so much," she commented.

"Put on a pair of loose jeans and you'll feel fine," her cousin suggested.

"Nothing I brought with me is loose. I'm doomed to go

back to California a two hundred pound wrestler."

"Don't be dramatic, Erin. You couldn't weigh an ounce over one fifty."

They both laughed.

"Maybe I can run it off tonight," she said quietly.

Neyse nodded. "We could make sure you do all seeking and no hiding."

Erin made a face. "That doesn't sound like much fun."

"Just a suggestion," Neyse replied. "I don't think one fifty looks that bad on you."

Erin nudged her, nearly knocking her off the tailgate.

"Be nice, or I won't let you play with us tonight."

"What are you going to tell your mom we're doing?"

"I'll tell her we're going for a walk. We are, after all. It's not a lie."

"No," Erin agreed, "not exactly."

"Nobody's going to get hurt and that's the only reason they worry."

Although it sounded convincing, Erin knew there was something wrong with the argument. She was aware of a vague discomfort but brushed it aside. *Neyse's right,* she reassured herself, *nobody's going to get hurt.*

Juan arrived as planned. He walked into the yard and up to the truck.

"Ready?" he asked.

"Ready," Neyse said with a nod. She turned toward the living room window and called out, "We're going for a walk, Mom," giving her mother little opportunity to say anything in reply.

Erin glanced back nervously as they hurried down the road toward the church. "Nobody's following us," she observed with relief.

"Relax," Juan told her. "We do this all the time."

"Okay," she said, not completely convinced.

Several young people were milling around in front of the church. Erin saw Alonzo immediately and then, as they drew closer, she saw Luis and Carlos as well. Her heart sank when she realized that Paolo was not with them.

"Hey everybody," Juan drew the group together as he and Neyse and Erin joined the rest, "you all know Neyse's cousin, Erin, right?"

Since Erin had already met many of the group, there was a general greeting.

"Who else is coming?" Juan asked.

"My brother said he'd play," one chubby girl replied.

"Felipe and Simon told me they were coming," Alonzo said.

"My sister went to get Natoma," another boy said.

"We can wait a while before we start," Juan announced.

Erin noticed that the others seemed to follow his leadership without question. He had a confident presence the others lacked. She wondered why, and made a mental note to ask Neyse later.

She saw Paolo, suddenly, walking from the direction of the plaza and she smiled.

Juan greeted him. "Thought maybe you weren't going to make it."

"No," he explained, "we didn't think it was a good idea to leave all at the same time."

There were seventeen of them in all. They set the boundaries: the canal to the east, the road by the *kivas* to the north, the main road west of the pueblo and the post office road to the south. The church fence entrance was base.

"Teams or singles?" Juan asked.

A discussion followed. They decided that they'd go in teams of two and drew straws to see who would be IT first; Alonzo and Natoma, a pretty, quiet girl, lost. The rest chose partners.

"Take off!" Alonzo cried.

"Come on, Erin," Neyse yelled, heading down the road toward the plaza.

Neyse ran and Erin followed, both girls being careful

not to make any noise as they passed by houses and slowing to a walk whenever they saw an adult. They hid in a small enclosure down the road from the plaza that had a clear view of the church. They struggled not to giggle at each other when Alonzo came within only a few feet of the spot.

"How are we ever going to tag in?" Erin whispered after he'd left.

"We'll just have to wait and see," Neyse replied.

Unfortunately, he stayed between them and base. They watched as several other teams got in free.

"We'd better make a move," Neyse told her at last.

They crouched, ready to take the chance.

Suddenly Neyse nudged Erin. They jumped up and raced toward base. Just then, Alonzo, who was about an equal distance from it, turned and saw them. He shot back toward the church.

"Run," Neyse yelled at Erin.

"You'd better be quiet," Erin puffed, "or everyone'll know what's going on."

"Run," Neyse repeated in a slightly lowered voice.

Alonzo beat them back. "Too bad," he said with a grin. "You two are the first ones we've caught."

Neyse moaned. "Guess we're next."

Before long, everyone else had been caught or tagged in and the game began again. Erin watched Juan and Paolo head off in the same direction as Carlos and Luis.

Neyse boosted herself up on the adobe wall. "We'll sit here for five minutes to give everyone time to hide and then we can take turns looking." She thought a minute. "Actually it makes more sense for you to stay near base since I know the best places to look."

When the time was up, Neyse started off on the road to the canal and Erin began pacing back and forth in front of the church. The moon was out, about half full, and the stars shown like tiny crystals in a black pond. She tried to listen for footsteps but could hear nothing but muffled music and a loud TV program. She turned slowly and gave a sharp cry: Paolo was leaning against the entrance gate.

"How...?"

He grinned, put up his hand in a mock greeting and repeated, "How."

She laughed. "Well, I told you I wanted to learn the language," she said, walking toward him.

"Looks as if you're a natural. Only one thing. You have a long way to go with hide and seek."

At that moment three others, including Juan, ran up and tagged in free.

"Wait a minute," Erin complained. "Interference."

"Sorry, Erin," Juan said shaking his head. "No such thing. You'd better get moving or you and Neyse'll be IT again."

She looked and she listened but she only managed to tag one girl. Luckily Neyse got another team, who found themselves tangled up with an unhappy goat, so they didn't have to be IT again.

"Let's trade partners," Juan suggested, tugging at Erin's arm. "Paolo can show you where to hide just as well as Neyse can."

"Probably better," Paolo said, taking her other hand and leading her toward the canal.

Her delight at his touch was short-lived for they had only gone a few yards when he let her hand fall.

"Where are we going?"

"You'll see. No need for impatience."

His teasing voice didn't have the sarcastic edge she'd heard so often.

When they reached the canal, Paolo started walking upstream. Erin followed as closely as she could, watching her step and battling the insects.

"I think I'm being eaten alive," Erin said. She brushed a mosquito from her arm, leaving a streak of blood.

Paolo took her hand and pulled her up a steep bank.

They found themselves behind a small house whose back yard overlooked the canal. The gates were open and Paolo led her in.

"Some friends live here. They're up north for the weekend."

Two beehive ovens, one smaller than the other, took up one corner of the yard.

"Here," Paolo said, indicating the four foot clearing between the two ovens.

Erin sat down self-consciously on the smooth ground, noting that there wasn't much room for two of them in the small space. He sat beside her, barely touching her leg.

"Now we can see both ways on the road but nobody can see us."

Erin looked down the road and realized that they were nowhere in sight of the church or any of the other young people.

"When did you discover this place?"

"Oh, a while back."

They sat. In the heavy silence, her ears began to ring. She shifted trying to dispel her discomfort.

"Do you have your cell with you?" he asked, his voice harsh.

She hesitated before she answered, "Yes."

He shook his head in disgust. "For heaven's sake, turn the stupid thing off!"

"I've got it on vibrate."

He gave her a look of utter disdain. "Do you *ever* turn it off?"

"Sometimes," she said, and then added, "Don't you have a cell?"

"Absolutely not," he replied fervently. He shook his head again. "I have no intention of selling out to Silicon Valley. People have survived without cell phones for centuries and now they need three or four in one family. What rot!"

"They're a convenience," she countered.

"They're an addiction," he scoffed. "How can I tell? Because people drop them in toilets. In toilets! They can't even go to the bathroom without them! Absolute addiction."

She sat silently trying to understand his venomous response, but she couldn't. So much anger about a phone made no sense to her.

"When are we going to go back?" she finally asked him.

"We aren't. We're going to let them look."

"But you said they wouldn't be able to find us."

"Well, we'll probably have to go back eventually."

"Oh."

It was quiet again for a while.

"What did you think of Taos?"

Her heart sank. *Let's not argue*, she pleaded silently. She hesitated.

"What's the matter? Don't tell me you didn't like it."

"I liked it."

"You certainly don't sound very enthusiastic."

"It was pretty crowded."

He laughed, deep in his throat so only she could have heard him.

"Too many touristas?"

"Yes."

The feeling that she'd had on the bridge came over her again: of watching the dreadful intrusion.

"Is it like that every day? Open to the public, I mean?"

"Every day but Christmas."

One holiday a year, she thought. *One day without the tourists*.

"Do you suppose that they mind it very much?" she asked in a whisper that almost didn't come out.

As he looked at her, she felt his eyes search her face in the half-light. Their eyes met and hers filled with tears.

"It was awful," she said quietly. "People were everywhere, looking at everything and everybody. They were noisy and pushy and, I mean, it isn't a museum after all, it's their home. People live there. Why can't they live in peace?"

His eyes never left hers. He frowned slightly, his head cocked to the side. When she'd finished, he reached out and gently wiped her cheek. He cupped his hand under her chin and tilted her head up. "Why the tears?"

She swallowed. "I hated being there. I didn't belong there. I had no right. Nobody invited me. It wasn't like being here."

"You went because you wanted to," he reminded her in a soft voice.

She nodded. "Maybe that's not good enough."

"The people there have agreed," he told her. His voice now sounded resigned. "It's an open pueblo. That's the price they pay."

"I'd hate to live like that, to be watched every day. It's as bad as a—a zoo."

"Erin," he said quietly, and then he didn't speak for a long time. "It's too bad other outsiders aren't as sensitive as that. We could take down all the signs."

He leaned over to her and put his arm around her shoulders. She closed her eyes and laid her head against his chest, feeling and hearing the rapid beat of his heart. She looked up to ask him a question but had no chance to do so. Paolo lowered his face to hers and touched her lips, gently, for what seemed like a wonderful forever.

Her senses spun in delight.

Paolo reached for her hand. Holding it, he traced each finger, and then closed it in his own. Erin felt as if she were floating. She hoped she would never come back to earth. He kissed her again, pressing his chest against hers, kissed her harder than before, yet without roughness. With what she could gather of her thoughts, she wondered what he was thinking.

Suddenly he stood and pulled her to her feet.

"Better go back," he said in a husky voice.

She was so stunned by his abruptness that she could scarcely move. He put his arms around her, bent, and kissed her lightly, almost like a father.

She didn't want the moment to pass. She was breathing almost as hard as he was and she knew, from times with boys back home, that the situation could lead to an argument. But this was different. For the first time in her life, the excitement was thrilling.

"Do you suppose they're still looking for us?" she asked.

"Our people don't give up easily," he replied, grinning at her.

He held her hand as they walked, but when they came into sight of the church he let it go with a squeeze. She wasn't surprised, but she was disappointed.

They snuck down the side of the community house, where they saw everyone else hanging around the front of the churchyard.

Juan saw them, looked around quickly and motioned for them to make a run for it. The frustrated home team saw them and beat them to the gate.

"Where have you been?" Neyse asked loudly. "They've looked everywhere for you."

"Never you mind, cousin," Paolo answered in a mysterious way.

"Well?" Neyse questioned Erin.

Erin shrugged her shoulders. "I don't know where we've been," she replied. "We went around in circles."

Neyse didn't look satisfied with the answer but she didn't ask anything else.

"Let's play again," Juan said. "Paolo, you and I can be IT," he half asked and half stated.

Paolo gave Erin a reluctant look. "Okay," he agreed.

"Come on, Erin," Neyse said. They ran off together toward the post office. As they rounded the corner, Neyse lost her balance, reached for Erin, but missed and fell.

"Agh...my leg!" she cried out, grabbing herself instinctively.

Erin dropped down beside her. "What's wrong?"

"I twisted my ankle!"

She cradled her foot in her hands, rocking back and forth, her face a study in fierce determination not to cry.

By this time the others had heard the commotion. Erin looked around helplessly, relieved to see the others coming. Juan and Paolo rushed up to Neyse, reaching her before any of the rest.

"What happened?" Juan asked.

"I tripped. Oh geez! My ankle hurts like crazy!"

Juan and Paolo exchanged glances.

"Can you move it?" Juan asked.

Neyse tried, cautiously. "A little," she said, wincing.

"Can you stand up?" Paolo asked.

"No. I don't know. I don't want to try."

"Come on, Neys. Let's see how bad it is," Juan coaxed.

Juan helped her up and she tried to put her foot down.

"I can't."

Juan shook his head. "I think I'll have to carry her."

"Mom and Dad are going to have a fit," Carlos said.

Neyse glowered at her brother. "Thanks."

Carlos looked apologetic. "Sorry."

"Don't worry about that now," Juan said. "Let's just get you home. Help me, Paolo, will you?"

Juan picked her up carefully and Paolo helped him balance while Juan adjusted her in his arms. Then, he carried her down the road toward home, her arms locked tightly around his neck. Paolo walked ahead of them, steering them away from uneven spots in the road. The rest drifted off slowly in other directions, mumbling quietly.

Erin and Carlos ran slightly in front. When they reached the house, Carlos threw open the door.

Tony and Julia sat together on the couch, watching

television. They both jumped up when Juan carried Neyse into the room.

"What happened?" Tony asked, nearly shouting.

Juan carefully put Neyse down on the couch, his face full of concern.

Neyse started crying. "It hurts, Daddy," she sobbed, finally giving vent to her pain.

As Juan told Tony about the accident, Julia propped the foot on a pillow on the couch and went into the kitchen. Erin heard ice cubes being dumped onto a towel. Julia applied the ice pack to the already swollen ankle and Neyse howled.

"It'll help, honey," her mother said encouragingly. "Give it a couple of minutes."

"Now," Tony said, "why were you running?"

Neyse looked unhappily at her father. "We were playing hide and seek."

"*What?*"

"Not now, Tony. We can discuss it later, okay? It's done," Julia said.

"You know better, Neyse," Tony said. "This is exactly why we don't like the game."

"Oh, Daddy. It could have happened just as easily in the middle of the day."

"But it didn't," he said, a mixture of anger and concern on his face.

He looked at the ankle, moved her foot carefully from side to side and watched her wince. He shook his head. "I don't think it's broken. There isn't much we can do for it tonight. Maybe a couple of Tylenol. You may have to stay off it tomorrow."

"Shall I get the Tylenol?" Erin asked, anxious to help and happy to get out of the way.

Julia nodded. "Thank you, dear."

"Do I have to go to the doctor?" Erin heard Neyse ask as she went in to the medicine cabinet.

"We'll see how it looks tomorrow," her mother said.

Neyse took the tablets and leaned back against the cushions.

"I think I'll get going," Paolo said. "Sorry, Neys." He patted his cousin's shoulder. "Hope you feel better."

"Thanks."

Erin walked outside with him. "Didn't turn out very well, did it?"

"She'll be fine," Paolo assured her. "Don't worry, okay?"

"Okay," she said.

"I mean it," he told her with emphasis.

She nodded.

He put his arms around her tentatively and leaned down close to her face. "I'll talk to you tomorrow," he said and kissed her tenderly.

She stood on the porch and watched him walk down the road, wishing she didn't have to go back in and talk to the others again.

She wanted to crawl silently into bed, go to sleep, and dream of the sweetest kisses she'd ever known.

CHAPTER 8

"Will I be able to dance for the feast?" Neyse asked, her voice tense with emotion.

"I seriously doubt it," the doctor said.

"But it's still almost two weeks away."

"Don't argue, Neyse," Julia said.

Neyse had been unable to put any weight on her swollen foot all day Sunday, and when Julia had called the doctor on Monday, he'd said to bring Neyse in so he could take a look at it. Erin had gone along to keep her cousin company and was surprised that the nurse let her into the examination room. She sat silently, watching the despair grow on Neyse's dark face.

"But…" Neyse began.

"I'm not trying to be mean, young lady," the doctor said. "Your foot will tell you the same thing when the time comes. You won't be able to make it through one dance, much less through an entire day of it."

Neyse lowered her head and looked mournfully at her swollen foot. Erin could feel her frustration.

The doctor continued. "I'm going to bandage it and I want to see you again in a week. You'll need to get crutches for at least that long. The less you're up, the more quickly it will heal."

Neyse didn't say another word as the doctor wrapped the injured ankle. When he was finished, Erin helped her hop out to the parking lot.

The front seat was already hot.

"Say something," Erin encouraged her, helping her get into the car.

Neyse shook her head sadly. "Me and my stupid ideas."

Erin shrugged. "You were right, you know, when you told your dad that it could have happened in the middle of the day."

"But it didn't," Neyse said, repeating her father's exact words.

Erin decided not to argue.

Julia came out of the building and got into the car.

"We'll have to go to Santa Fe to get crutches," she said, "but first I need to let Carlos know we'll be gone for a while."

They drove home in silence.

As they pulled up in front of the house, Carlos came out to the front porch. "What did the doctor say?" he asked.

Julia leaned over the front seat so she could see past the girls. "It's only a sprain, but we need to go to Santa Fe for crutches. Do you want to come with us?"

"No, thanks. I'm going to meet Luis at the rec center."

"All right. It shouldn't take us too long."

She blew him a kiss.

"See you later."

"Oh, Neyse," Carlos said. "Paolo came by while you were gone. He wanted to know how you were doing. He said to tell you he'd come by again later." He darted back into the house.

"That's weird," Neyse commented.

Erin smiled inwardly but said nothing.

By early afternoon they were back home, Neyse

propped up on pillows on her bed, resting the now bandaged ankle. They talked about the weekend and the fateful game on Saturday night.

Neyse's cell rang. She answered and after a couple of seconds, got a strange look on her face. She talked briefly, thanked the caller and hung up.

"That was Paolo. He wanted to know how I was doing. He says he's coming over later to see me. He says he hopes I'm not in too much pain."

Suddenly Neyse looked straight at her cousin and without warning said, "Erin, what the heck happened with you and Paolo the other night? I've been hurt before and he's never been particularly concerned about me."

Erin hesitated. "Well...we talked some...and..."

"And?"

"Nothing much. We talked, that's all. I wanted to get to know him better, you know, so he wouldn't be nasty to me so much."

"Guess it worked."

"He seems different from guys at home."

"He's different, all right."

Erin curled up in the corner of the lower bunk, next to the wall. "Tell me about him," she coaxed.

"Well, he's a loner, for one thing. Other than Juan, he doesn't have many friends. Like I told you the other day, he's pretty radical about the rights of Indians. He has a real thing about it. He thinks that the Indians are the chosen people. He thinks we've been taken advantage of for centuries and that the only way to preserve our culture is to keep the race pure."

Erin looked at Neyse carefully. "How does that make you feel?"

Neyse shrugged. "It doesn't bother me. He's always been good to me, and he seems to really love my mom. His opinion of Anglos is limited. He doesn't really even know any. There were Caucasians at school, of course, but he avoided them entirely."

"Does he have a girlfriend?" She asked it, not at all sure she was ready to hear the answer.

Neyse thought a moment. "I don't think so. He's gone out with Natoma a few times. Other than that, I don't think I've ever seen him with anyone. Boys here are careful not to show much attention to a girl unless they're really serious. They'd get teased too much."

Erin looked out the window with a wistful stare.

"What's going on?" Neyse shouted in a whisper.

"Nothing is going on. It's just that—well, I like him. And I think that maybe he likes me, too."

"Did he say that?"

"No, not exactly. He just sort of acted like it."

"Well, coming over to see about my foot is definitely a sign of something. You know, Erin, it wouldn't be easy for him to admit that he liked an Anglo."

"What?"

"What I mean is, don't let yourself get too interested, or you might get hurt."

"How do I not get too interested?"

"I don't know. I just don't want you to be disappointed."

"I'll do my best to get uninterested, then," Erin said, grinning.

Neyse's cell phone rang. "Its Juan," she said, looking at the number. She talked to him briefly, smiling and even laughing a little.

"Hope you don't mind, but I just made plans for us for tonight."

"What kind of plans?" Erin asked.

Neyse eased herself back onto the bed. "Juan talked to Paolo today while they were up at the rec center and they decided to take you on a tour of the area around the pueblo. They'll be here to pick us up around six o'clock."

"Seriously? Paolo, too?"

"Yes."

Erin stopped suddenly. "Will your dad let you go out after the other night?"

Neyse looked down at her ankle. "He thinks that not being able to dance is the worst punishment possible. In a way, he's right," she added quietly.

Erin couldn't help but think that if she'd broken the rules at home, she would have been grounded for weeks.

Erin had never known two hours to pass so slowly. She changed her clothes three times, trying to look nice, but not too nice. She didn't want Paolo to be put off. She ended up wearing denim shorts with a bright lavender cotton peasant blouse of Neyse's.

She tried her hair in braids, pig tails, a pony tail, pulled it up on the sides and decided, finally, to simply let it hang down straight.

She laughed when she emerged from the bathroom; Neyse had tied four blue ribbons around her bandaged ankle, the color matching her shirt exactly.

"I want to look pretty, too," she said with an impish grin.

Erin was glad that she and Neyse had talked about Paolo. She'd been uncomfortable about hiding her feelings for him. The conversation earlier made getting ready for their date much more fun.

"Are you ready?" Neyse asked. "Or do you think you need to try on the rest of my clothes?"

Erin looked down at herself. "I'm ready. Do I look all right?" she added uncertainly.

"You look terrific."

When they heard the muffled rumbling of a car in the driveway they hurried into the living room. Erin peeked out the window, turned and made a face.

"It's your dad." she said, and joined Neyse on the couch where they sat like two wallflowers at a dance.

Julia came in from the kitchen. "You two certainly are

acting strange. What's going on?"

"We're just waiting," Neyse said.

"Don't you think you should eat something before you leave?"

"We had lunch late, Mom. I'm not even hungry."

"I'm not either," Erin seconded. Her stomach was churning so much that she couldn't have eaten if she'd been starving.

The front door opened and Tony entered, a weary smile on his face.

"How nice of you to greet me like this," he said.

The girls smiled at each other.

"I'm sorry to disappoint you, dear," Julia said, putting her arm around him, "but they're waiting for Juan and Paolo."

"We're going to go play hide and seek," Neyse said, her look assuring him it was the last thing on earth she would do.

Tony started to reply but his wife's hand on his arm and the sound of a car engine stopped him. Erin went to the front window, peeked out and turned, smiling.

"Have fun," Julia said as the two girls stepped out onto the front porch.

"We will," Erin called back.

They smiled at Juan and Paolo.

"Good evening, ladies," Juan said grandly, as he and Paolo got out of the car. "Your limousine has arrived."

Paolo looked as if he'd stepped out of the shower five minutes earlier, his hair still wet, the collar of his shirt damp.

Erin helped Neyse navigate out to the car. Juan had thrown several pillows on the floor in the front seat so that Neyse could rest her leg comfortably.

"I feel like Cinderella, with all this fuss over my foot."

"You're worth it, Cindy," Juan said, patting her leg.

After she had gotten settled, Erin slid into the back seat. Paolo got in and sat beside her—not too close, she noticed.

"Where's Tiya?" she asked as Juan started the car.

"He had to stay home tonight," Juan said before Paolo had a chance to speak. "He was expecting a visit from a lady friend."

Paolo laughed.

"Where are we going?" Neyse asked.

"Just sit back and relax and enjoy the view," Juan told her.

"That's not much of an answer."

Juan headed south, in the opposite direction of the lake. He began to talk, apparently to Erin. "The pueblo closest to ours, both in distance and custom, is far more conservative in its traditions. It's said that we were originally one village but split in two for some reason."

"What do you mean by more conservative?" Erin asked.

"Despite what's happening in the outside world, they continue to teach and stress the importance of the pueblo religion and traditions. The children learn our language from birth and they speak it in the family almost exclusively. Few people have phones or televisions and influence from the Anglo world is discouraged. Marriage with anyone from outside the pueblo is rare."

Erin listened, looking at Paolo when Juan finished. "You must admire that."

"It's the only way to preserve the culture. Even at that, it's a struggle. Whites assume that their culture is superior to ours. There's no room for an alternative one. The only way they see for us to be successful is to change—to live their way. To the extent that we do that, we let go of our own traditions and they're gone forever."

His voice was sad, not angry as it had been when he talked about the influence of the Catholic Church the day she'd arrived.

"It's strange. Whites have groups to save the whales, to save the condors, to save just about every endangered species. Indians, whose numbers decrease every decade, are on their own."

The thought that the Pueblo Indians could lose their individuality—their very identity—seemed entirely possible. History had many examples of population assimilation. Even Scott talked about the society making "carbon copy people".

"The white man isn't the enemy," Juan said, "and it's possible we wouldn't have survived this long without his technology. The choice is ours, really, to accept what we want and reject the rest. We don't need to fight."

Erin felt Paolo stiffen. He didn't have to say a word for her to know he disagreed. He turned his head and looked out the window.

Erin could not imagine how she'd neglected to turn off her phone, but at that completely inopportune moment, it rang.

Paolo spun around and growled, "Oh, for heaven's sake, do you have to take that thing everywhere?"

"Lighten up, Paolo," Juan said as Erin pulled the phone out of her pocket and silenced it.

"It's like a noisy appendage," Paolo complained. "Nobody seems to be able to make a move without one," he added with disdain.

"We all know *you* do," Neyse chided.

"Come on. Nobody seriously needs a cell phone."

"I'm sure having one has saved more than one life," Erin argued in her own defense.

"Could we please not revisit this topic?" Neyse pleaded.

Paolo simply stared at the view.

Erin wished that Juan would turn on the radio.

The river ran along the side of the road, cottonwood leaves shimmering in the late afternoon sun. After a time, pavement turned into well-worn dirt roads and Erin again saw rich brown adobe buildings, smaller and older looking than those at home. Children ran and played, shouting to each other words Erin could not understand.

Juan slowed the car to a crawl, allowing Erin to see more easily.

"The plaza here is much larger than ours," Juan said, "rectangular, rather than square."

Erin did her best to see it through the narrow passageways between houses. All she could make out was an open area, surrounded by adobes built as closely together as possible.

"How many people live here?" Erin asked.

"About three times as many as in our village," Paolo told her. "That's one of the advantages of the strict adherence to tradition—people don't leave."

Erin didn't hear the defensive overtone she'd heard in his voice other times he'd spoken of maintaining a strong Indian identity.

"The disadvantage," Juan said, "is people who are frustrated because they can't break out of the mold created for them by their ancestors, a mold that may no longer suit their needs."

Erin waited for a barrage of arguments from Paolo but they didn't come.

Juan continued to talk, sounding almost as if he were giving a classroom lecture. "Our pueblo tradition is based on the fundamental religious belief that each person must live in absolute harmony with the natural world. Every village is built around a central plaza where we have dances and ceremonies to celebrate our philosophy."

"The dances are sort of like church services," Neyse said, simplifying Juan's information.

"Outsiders come to watch the dances," Paolo said, "making no effort to learn anything about it beforehand. They talk loudly, take pictures, even clap. They wouldn't think of doing that during one of their church services."

"At least they are interested enough to come," Juan said. "Maybe they'll try to find out more about our culture afterward. Our best chance for survival is through the understanding of the larger society."

He slowly drove through the narrow streets, careful to

avoid the many children playing there. The four friends received little attention from the village children and only mild looks of curiosity from the adults who saw them.

Knowing how strict this pueblo was, Erin felt like an unwanted intruder and was relieved when they had left the village behind.

"Let's go up to the old village," Neyse said as they started back toward home.

Juan laughed. "That's a bit of a walk for a girl with a broken leg."

"My leg is *not* broken," Neyse wailed. "Besides, all I meant was to drive up to the mesa."

"That's exactly where we're going," Juan told her.

They drove back, past their pueblo, past the lake and then further north. The dirt road was rough, full of rocks and potholes. It led them into a mountainous area, prettier, Erin thought, since there were trees and bushes breaking up the harsh landscape. They bounced along, watching the sun setting slowly on the far horizon.

About a half an hour after they'd passed the lake they came to a fork in the road. Juan took the one to the left, drove another mile or so, and then pulled to the far right-hand side of the road.

"The old village was on top of the mesa," Paolo said, pointing to a huge, red mountain bluff that rose some good distance from the car. It looked as if the top of a cliff had been sliced off with a giant knife, leaving a flat plateau. His mood had lifted, which was a huge relief to Erin.

"Table Top Mountain," Erin said, giving a rough translation of the Spanish word *mesa.*

"How many years of Spanish have you had?" Juan asked.

"Three." She looked back at the mountain. "How did they ever get up there?"

"There are several ways, some easier than others."

"It's wonderful up there," Neyse said. "I wish I hadn't

done this," she added, pointing to her foot, "so we could show it to you."

"I could show it to her," Paolo said.

"Oh, Paolo, that's a great idea!" Neyse cried.

Erin looked back up at the mesa. "Now?"

Paolo laughed. "Hardly. It takes hours to hike up and back. Sometime after the feast would be better."

Erin wasn't sure if he was hesitating or not. Although hiking wasn't her favorite thing, she was curious about the ruins and would agree to go just to be with him. She waited, hoping that he would suggest a particular day, but he didn't.

"Well," Neyse said, "I can go to Dixon's."

"Dixon's?" Erin repeated. "That doesn't sound very Indian."

"Good call," Juan said chuckling, and he turned the car around, drove back to where the road had split and took the fork to the right. Shortly, a valley appeared before them, lush and verdant, with a carpet of green apple trees. Erin hadn't seen anything like it since she'd left home.

"Every fall we come up here and get crates of apples," Neyse said. "It smells heavenly when the fruit is ripe. Mom cans applesauce and makes pies to freeze and bake later."

The sight of the greenery was completely unexpected; Erin imagined that a farm had somehow been magically dropped in the middle of the surrounding desolation.

"The Dixon family has owned this place for ages," Juan said. He drove past the last of the rows of trees, across a small bridge, and headed toward the large, white house.

"It's beautiful," Erin agreed. There was even a lawn in front of the house. She wondered where the water came from to keep it all so green, but decided not to ask, afraid it might bring up a sore point.

"If it weren't for me," Neyse complained, "we could get out and walk around,"

Juan patted her leg. "Don't blame yourself. I know that Erin tripped you."

"What?" Erin exclaimed.

"No apologies," Juan said. "It's all over now. Let's try to forget it."

Paolo laughed.

"I didn't..." Erin began.

Paolo put his arm around her shoulder. "Don't deny it. We know the truth."

Erin moved a little closer to him and he squeezed her arm. Her heart soared.

"See the barn?" Neyse asked, pointing past the farmhouse. "That's where they sell the apples. Mom and Victoria really do buy them by the crate."

"All this talk about food is making me hungry," Juan announced. "Let's go get an ice cream."

The others agreed readily.

As they drove back, Neyse's favorite song came on the radio. She turned up the volume and sang with the group.

Paolo leaned close to Erin. "Do you really want to hike up to the ruins?"

"Yes."

"I won't have any time off work until after the feast."

"That's okay." After a pause she added, "I'm not much of a hiker."

"Don't worry. We'll take our time."

He said it just loudly enough for her to hear, his voice reassuring and seductive.

Almost before she knew what had happened, they were sitting in front of the small village market.

"I think I'll wait here," Neyse said as Juan opened the car door. "Just get me a fudge bar, okay?"

Juan nodded.

"I'll keep you company," Erin said.

"What do you want?" Paolo asked her.

"A fudge bar sounds good."

Paolo and Juan went into the store. The lights spilled out onto the parking lot, throwing soft shadows on the asphalt. Inside, Paolo held up a frozen chicken and asked, by gestures, if that was what they wanted.

Erin laughed, shaking her head no.

"Yes, yes!" Neyse yelled, but he pretended he couldn't understand her.

Paolo took the chicken toward the cash register as if to buy it. Juan walked by the window with two bananas on his head and Paolo pantomimed shooting them off with a bow and arrow.

"No," Neyse cried out. "A fudge bar."

Erin howled. *What the heck happened? He's like a different person.*

They came back to the car at last, got in, and gave the girls each their fudge bars.

"*Dawaa'e*," Neyse said. She looked at Erin expectantly.

Erin's eyebrows went up. "Thank you?" she said, guessing the meaning of the word.

"*Ha'a*," Neyse said, smiling.

Erin opened the package of ice cream and looked inside. She frowned. "I asked for a fresh peach cone."

Paolo made a face at her.

"Did you have to say that?" Neyse asked. "It's my favorite."

"Mine, too!" Erin cried. "In fact, we made peach ice cream for my birthday last year."

"When is your birthday?" Juan asked.

"August nineteenth, just a day after Neyse's."

"Let's have a party together and make peach ice cream," Neyse said, her voice shrill with excitement.

"Are we invited?" Juan asked.

"What kind of a party would it be without you two?" Neyse asked.

A lousy one, Erin answered silently.

"Good, then it's settled," Juan declared.

By the time they arrived at Neyse's front door, they'd all finished their ice cream bars. Neyse was helped into the house and guided to the couch.

"Have fun?" Julia asked, coming into the living room.

"Yes!" Erin said.

"That's good. Anyone want dessert?"

Neyse laughed. "We just stopped for ice cream."

"That doesn't mean we wouldn't eat apple pie, if you had any," Juan told her.

Julia smiled at him and shook her head. "Sorry."

Juan shrugged. "It was worth trying."

"I'm going to go read for a bit," she said. "Good night, all."

It was almost midnight before Paolo and Juan left. Paolo pulled Erin gently out to the front porch while Juan told Neyse goodnight.

"Don't forget about the ruins," he said.

She shook her head. "I won't." She wondered if he was saying he wouldn't see her before then.

He leaned down and kissed her. "I had fun tonight."

"Me, too."

"I'll see you," he said, his voice trailing into another kiss.

Juan opened the screen and Paolo instantly pulled away from her.

"I'll see you," she repeated, relieved Juan apparently hadn't noticed them kissing.

She watched as they backed the car out and drove down the road.

"Soon, I hope," she added to herself.

CHAPTER 9

"Oh, *no!*"

The voice came to Erin through a misty gray fog.

"Get away. Go on. Get *out* of here!"

Her eyes opened slowly, her mind resisting. She sat up in bed, recognizing Julia's angry voice.

"Go on! *Go on!*" her aunt yelled.

Erin jumped down out of the top bunk and shook Neyse. "Wake up! Something's wrong."

Neyse rolled over, listened for a moment and said sleepily, "Cows."

"Cows?" Erin repeated, not sure she'd understood her cousin correctly.

"Cows," Neyse said again. Pulling on her yellow and white cotton robe, she hopped out to the front porch, Erin following close behind.

"Out! Out!" Julia cried.

The girls began laughing when they saw Julia trying to push a stubborn brown and white cow out of the yard, her long, pale pink robe billowing like a matador's cape.

"Daddy!" Neyse called back into the house.

Tony came running out of the house and hurried to help

Julia. It took the strength of both adults to force the animal out of the yard. Tony closed and secured the gate when the intruder had been ousted. Julia walked around to the back yard without saying a word. The girls exchanged puzzled glances. Tony shook his head as he passed by them, following his unhappy wife.

"What's going on?" Erin asked, as she and Neyse joined Tony and Julia on the back patio.

Tears edged the corners of Julia's eyes. "The dumb thing ate all my apricots," she said. "I won't have enough left to make even one pie. Why can't people keep their animals locked up?"

Tony put a comforting arm around her waist as she stood looking at the crate of apricots a neighbor had brought over the night before. Since there had been a few fruit flies on them, Julia had put the crate on the patio table just outside the kitchen door.

"It's not that bad," Tony encouraged her, but his tone lacked conviction.

Fruit was strewn on the patio floor, and much of what was left in the box was covered with a layer of the cow's slobber.

"You fix some coffee and I'll clean up the mess," he said, turning her around and gently pushing her toward the kitchen door.

Erin opened the screen, allowing Julia and Neyse to enter ahead of her.

Julia measured water into the coffee pot, mumbling about cows and food and irresponsibility.

"Mom hates the cows coming into the yard," Neyse whispered as she and Erin went back into the bedroom. "It's been better lately because people really have been trying to keep them penned up, but she says it's a waste to struggle to grow food and then have cows wander around freely and munch out on anything they can find. The funny thing is, they'll eat anything. One ate her only rosebush last year, and my mom nearly killed it."

Erin felt badly for Julia, but she couldn't help giggling. "She sure did look funny."

Neyse nodded in agreement as she closed the bedroom door.

"How's your foot?"

Neyse wiggled her toes. "Okay. It doesn't hurt much but I don't think I could walk very far on it." She tried to put her weight down on it and frowned. "The doctor was right. I'd never be able to dance."

"Is that bad?"

"Bad?"

"Will you get into trouble?"

"No. Everyone has the choice of dancing or not dancing. It's an honor. I love it. Besides, I wanted you to see me." Disappointment settled on her dark face.

"Can we go watch anyway?"

"Of course. You'll be able to see Daddy, and Carlos, and Juan..."

"And Paolo," Erin interrupted with a smile.

Neyse brightened at Erin's obvious pleasure. "Yes, and most importantly, Paolo."

"Are you girls getting dressed?" Julia called in from the kitchen. "I'm going to need a lot of help today."

"We're coming," Neyse called back, adding quietly, "We'd better hurry. Mom gets tense around Feast Day."

They dressed, joined the rest of the family, and ate a quick breakfast of fruit and cereal. Tony asked Carlos if he wanted to go into town with him.

"Sure. Want to go with us, Erin?" Carlos asked, hopefully.

"She can't," Neyse answered for her. "She has to help cook."

"Next time," Erin said.

"You always tell me that," he complained.

Erin went out to the porch to wave goodbye to her

young cousin when he and Tony left, feeling guilty that she'd hurt his feelings. "Next time," she called again as they pulled away, "I promise."

She went back into the kitchen and discovered Julia and Neyse already hard at work.

"What are you going to make today?" Erin asked.

"*We*," Julia emphasized, "are going to make pies, stews, beans, and salads. We'll start with the pies and preparation for the stews. Some things, like the bread, we can't even begin until the day before."

The girls cleaned and cut up prunes and cherries, and then the apricots Tony had managed to salvage. After removing the pits, they put each kind of fruit into a different cooking pot.

Julia made pie dough. "We'll make the fillings today and put all but the prunes in the crusts and freeze them. That way I can cook them all at the last minute and they'll still taste fresh. The prune filling will go in the refrigerator and I'll make those the day before the feast."

The house filled with the sweet smell of stewed fruit. Erin watched Julia ladle steaming apricots and cherries into thick-edged pie shells. She put the prunes into large plastic containers.

"I'm starving," Neyse announced, scooping a finger full of cherry filling from the side of an empty pot.

Her mother looked at the clock on the stove. "I'm not surprised. It's almost two. How about some cantaloupe and cottage cheese?"

"Sounds yummy," Neyse said.

Julia made an attractive fruit salad for each of them and they ate together, glad for the rest after the morning's activities. As the heat of the day pressed in on the house, the air conditioner droned overhead.

"I hope it won't be too hot for the dance," Neyse said.

Erin looked amused. "How hot is too hot?"

"Over one hundred."

"I'm sure it's more than that now," Julia commented, her face flushed and damp.

"We should go swimming at the rec center and cool off," Neyse suggested. "Would you take us up there, Mom?"

"Since you were so much help, I think I can manage to do that for you. I wonder if Paolo is working today," she said, smiling at Erin.

"You could go with us and see," Erin teased.

"Thank you, but no. I'm not much of a swimmer."

The girls washed their dishes and hurried to change into their bathing suits.

Erin stripped off all her clothes and as she was pulling on her suit, Neyse said, "Your tan is getting pretty dark."

"I told you I'd be able to pass as your sister by the end of the summer."

"Please notice that as your body gets darker, your hair gets lighter."

Erin looked in the mirror. Her hair was streaked, almost white in places.

"I'll wear a hat."

"You'll need to wear a bag," Neyse said, giggling.

"Thanks a lot."

"I meant that you don't look very Indian. You look good, just not very Indian."

Erin's one-piece, bright turquoise suit was a perfect cut for her slim body, raised high over the thighs and low in the back.

"Paolo may have trouble concentrating on his job when he sees you in that."

"He's seen me in it before...at the lake."

"He probably wasn't looking at you in the same way he has been lately."

Neyse put on a pale yellow bikini that made her skin look even darker.

"Too bad Juan won't be there," Erin said. "It's a shame to waste that suit on your cousin."

Neyse put her arm around Erin's waist. "We do look good, don't you think?"

"Fantastic."

They grabbed towels, sandals, and sun block, and told Julia that they were ready.

"That didn't take long," she said.

Julia drove them to the rec center, all the windows in the car wide open. "I can't believe the heat."

"Mom, it's this hot every summer."

"I know. I guess I forget."

"You need to get the air conditioning fixed," Neyse commented.

Julia shrugged. Erin got the impression that her aunt agreed even though she didn't say anything.

Julia's face was red and perspiring. Erin was hot but it didn't seem to bother her like it bothered Julia.

"Why don't you go swimming with us, Aunt Julia?"

"No, I'm going to take a cool shower, and read for a while."

"Would you mind picking us up at about six o'clock?"

"Okay. But be ready so I don't have to come in after you."

The girls agreed.

Julia dropped them at the front door of the center and drove off, her hair ruffled by the wind.

They walked slowly into the center and down the narrow passage to the pool area. Neyse was no longer using crutches, but she was still babying her ankle. Enclosed on all sides by an adobe wall, the pool was surrounded by bright green indoor/outdoor carpeting, a strange sight to Erin. She looked down at it with a frown.

"If it wasn't covered," Neyse explained, "the deck would burn our feet right off."

The pool was good sized. Erin was surprised how few people were there considering how hot it was. About a dozen children were in the water, yelling and splashing. Erin spied

Paolo sitting on the deck at the side of the pool, dangling his legs in the water and talking to Alonzo.

Neyse threw her towel and sandals aside and eased herself into the water. Paolo waved to Erin and motioned to her to come over. Neyse beat her to him.

"You on duty?" Neyse asked.

"Until four o'clock."

Erin sat down next to him, but not too close. She knew that he would not appreciate any expression of intimacy in public. She let her legs hang over the edge of the pool.

"*Kuwe tzi*," she said.

He smiled his approval and rubbed his foot against her leg. "Hi." He leaned his head back and studied her for a moment. "No phone?"

She shook her head. "Not even *I* take a cell phone in the water."

"Please don't start again," Neyse pleaded.

"I'm here, too," Alonzo said in a loud voice.

Erin grinned at him, noticing how much he looked like Paolo when he smiled. "Hi," she said. "How are you?"

"Better, now, thank you."

Erin watched him as he swam off toward the deep end. Neyse had said that Alonzo was sixteen. He was good looking, without a doubt, but he didn't yet have the charm that his older brother had.

Neyse was bobbing up and down in the water.

"Come on in, Erin. If he has to guard until four, he won't be any fun until then."

"I probably won't be any fun then either," Paolo said, but his grin said otherwise.

Erin slid over the edge into the water and with a kick that she knew would drench him, swam away.

Alonzo was diving off the board, doing front flips and back flips and trying to convince Neyse how easy they were.

"You try, Erin. You must be able to do a flip."

She couldn't do much on a diving board but Scott had

taught her to do flips. She decided it would be fun to give Alonzo a bad time.

"I could never do a flip," she said. "I'm too scared to even try."

"Oh, come on, Erin. I'll teach you."

"Well..."

"Come on."

"Okay."

She climbed out of the pool and following Alonzo's instructions, stood on the end of the board with her back to the pool. Then, as she listened intently to what he was telling her, she teetered, gave a sharp cry and fell into the water.

"Balance," he was yelling as she came to the surface. "You have to balance while you get ready."

Back on the end of the board again she could hardly conceal her laughter. Her toes gripped the rough surface, her arms extended in front of her and she frowned in deep concentration. She threw her arms back and over her head just as he directed, made no attempt to flip, but merely jumped backwards into the water.

Alonzo was yelling again when she surfaced. Neyse was laughing gleefully. She turned to look at Paolo as she climbed up the ladder and saw on his face a mildly tolerant smile.

"One more time," she said to Alonzo. "If I can't do it this time, I'm going to quit."

"You can't quit. It could take ten times before you get it right."

"No. Just one more try."

She stood, balanced on the board, listening to him. She threw her arms back again, as he told her, and her body as well, and landed flat on her back in the water. It stung as if she'd been hit with a board, and she wasn't sure it was all worth it.

"Are you all right?" Alonzo asked. He'd jumped in and swam out to her.

"Yes. I just need to rest for a while."

He hurried ahead of her and spread her towel on a lounge chair near Paolo. Neyse limped over and sat beside her.

"Your back is pretty red."

"It hurts just a bit," she said with a grimace.

"You put on quite a show," Neyse added, a knowing smile on her face.

"You can still do it, Erin, if you'll try a few more times," Alonzo encouraged her.

"Never again, Alonzo."

He sat with them for a while, but finally went back into the water.

Paolo turned to Erin and with a sly smile said, "Let me know when you're going to do the finale. I don't want to miss it."

She felt a little guilty then, that perhaps she was being unfair to Alonzo. But Neyse noticed her mood and poked her in the ribs. "No big deal. It'll be funny."

The afternoon lifeguard came on duty and after talking with him for a few minutes, Paolo asked Erin if she wanted to swim.

"Sure."

He made a low racing dive into the pool and she followed. They swam laps for about five minutes, Erin keeping pace with him easily. Paolo rolled and floated on his back for a while and Erin did the same.

"You know, I haven't forgotten that you drenched me."

"I thought you might like to get cooled off."

"I'm going to repay you for your kindness."

"No need," she said politely.

"It won't be a bother. Really. In fact, I'm looking forward to it."

She turned over and started to swim away but he grabbed her leg and pulled her to him. She squirmed but she couldn't break his hold on her.

"It wouldn't look good for a lifeguard to drown a swimmer," she sputtered, still trying to break away from him.

"That's true. I'd probably lose my job and it might ruin my summer."

"Your summer!"

He laughed and relaxed his grip. She took advantage of his momentary lapse, jerked away, and dove down to the bottom of the pool. He followed and when she saw him after her, she couldn't keep herself from giggling. She came up in a flurry of bubbles. She shot for the far end of the pool and scrambled up on the deck. She hopped up onto the diving board, grinning at Paolo, who by now was on the edge of the pool.

"Alonzo," he called to his brother. "Watch this. I think she can do it. I've got her hypnotized."

Erin turned around with her back to the pool, extended her arms, balanced on her toes on the end of the board and jumped, throwing herself up and back into a tuck position, flipping easily.

Alonzo ran to the board and was yelling at her, excitedly. "That was perfect. Just beautiful. How did you ever...?"

Paolo, Neyse and Erin were laughing together and Alonzo realized he'd been taken.

"Thanks a lot. Go ahead, make a fool of me. Get a good laugh." He stomped off.

"Alonzo, wait," Erin chased after him. She caught his arm. "Come on, be a sport. It was only a joke. It hurt me, too. My back may never be the same."

"Serves you right."

"You sound like Paolo," she whispered.

He grinned. "What a putdown."

"Will you forgive me?"

"Are you really a diver?"

"Heavens, no. Scott taught me to do flips, that's all."

"How long did it take him?"

"An entire summer."

"Serves you right."

They laughed together, the tension melting away.

For the rest of the afternoon, they alternated between lying beside the pool and jumping in the water to cool off, the only way to really get any relief from the heat.

Neyse had gone to get a drink and when she came back she said, "It's almost six."

"Oh, nuts," Erin moaned.

Neyse looked at Paolo and Alonzo. "You guys want a ride? Mom is coming to get us."

"Oh, you have a chauffeur service now?" Paolo teased her.

"What, you expected me to walk?" she returned, pointing at her foot.

"How long are you going to keep using that old excuse?"

She shot him a dirty look.

Just then, a horn honked and they saw Julia through the iron gate that led to the parking lot.

"Shall I carry you to your coach?" Paolo offered.

"I can make it, thanks."

Alonzo, Paolo, and Erin crowded into the small back seat, and Neyse sat in the front. Paolo put his arm up on the seat behind Erin, resting it lightly on her shoulders. His skin was warm against her, but even as hot as it was, she would have happily stayed that way for hours. Too soon, they were home.

"Want a Coke?" Neyse offered.

Paolo accepted as Erin hoped he would. Alonzo wanted to go fishing, so he declined.

Erin dropped ice cubes into blue plastic glasses and Neyse poured them each full of Coke. They finished off the cool drinks quickly.

"I'm going to take a shower and get changed," Neyse

said, winking at Erin. She disappeared down the hall.

"I suppose I should go," Paolo said, to her great disappointment.

"Would you like to come over after dinner?" As soon as she'd said it, she wondered if she'd made a mistake.

A strange look came over Paolo's face and he hesitated. Finally he said, "I can't. I have plans."

He sounded curt, almost as if she'd backed him up against the wall. Her heart sank.

He has a date.

He left quickly and she was sure of it. She worried about it all through dinner, feeling miserable and hardly eating a thing.

Later, as the family sat in the living room watching TV, Neyse called to ask Victoria a question and as she was speaking to her aunt on the phone, Antonio opened the front door.

"Where are you off to, Uncle Tony?"

He gave his niece a very long, very hard look and she could see that he didn't want to answer. "I have something to do," he said at last, and left, closing the door quickly.

Erin felt tears build in her eyes, wondering what she had done. What was wrong? She got up and went quickly into the bathroom, splashing her face with cold water.

Julia came in and put her arm around her. "Don't be upset. There are things, religious things, which are private. They won't talk about them to outsiders. It is strictly forbidden. He cares for you very much or he wouldn't have said anything at all. He went to practice for the dance. He doesn't even tell me. I just know."

"What if you had asked him?"

"I wouldn't have."

"But if you had. Aren't you curious?"

"No. It doesn't concern me. It has nothing to do with me."

"But it does have to do with you. You're his wife."

"The pueblo says it has nothing to do with me, so it has nothing to do with me. That's just the way it is."

"Would Neyse have gone, if she were dancing?"

"Yes, though some young people don't practice until just before the feast. Many go every night right along with the adults."

Erin thought about Paolo's strange reaction that afternoon and realized that was why he wouldn't answer her. She threw her arms around Julia.

"Thank you," she said, hugging her aunt with relief.

"You're welcome," Julia said, looking confused.

That night after she was in bed, Erin wondered if she could live like Julia; if she could be a pueblo woman and yet not. She wondered how long it had taken for her aunt to learn not to ask, to accept without questioning.

It was so foreign to her upbringing. *I don't think I could, even if I really loved someone.*

CHAPTER 10

The morning sun, hidden behind a thick cloud cover far to the east, struggled to break through and warm the day. Everyone had gotten up early, as the preparations for the feast would take every available hour.

"Thank goodness it's still cool," Julia said as she came inside through the kitchen door. She pushed it open, flat against the wall, with her leg. She was carrying a large tub that she placed on the kitchen counter.

"I thought we were going to make bread," Erin said, frowning at her aunt.

"We are."

"In a wash tub?"

"We're going to make *lots* of bread."

Neyse had heard the exchange from her bedroom where she was dressing. "You won't believe the crowd that will be here tomorrow," she called. "We'll *need* lots of bread."

Erin watched Julia pour a twenty-five pound bag of flour into the large steel-gray tub and mix in salt and shortening. Then she added a mixture of yeast and water.

Neyse came into the kitchen, still limping almost imperceptibly. Julia looked up and said, "Dig right in."

With obvious familiarity, Neyse plunged both hands into the tub. Erin tentatively followed her example. For several minutes they mixed the ingredients together with sticky hands, until, after a while, the dough began to take shape. That was when Julia took over, turning the dough and folding it into itself, finally making it into one gigantic lump. She put it on the counter and kneaded it for several minutes before replacing it in the tub that Neyse had greased. Then she spread the dough evenly and covered it with waxed paper.

"Will you help me put it out on the porch?" she asked the girls.

The three of them maneuvered the heavy, cumbersome load through the door to the patio table. Julia covered the tub completely with a faded old quilt and two blankets.

"Now we let it rise. Later we'll punch it down," she explained to Erin.

"That's the fun part," Neyse said.

Julia took several packages of beef out of the freezer. "I'd like you girls to cut up the meat for me—half of it in small cubes for the chili stew and the rest in larger pieces for the posole."

"What's posole?" Erin asked.

"A sort of soup," Neyse told her, "with hominy—that's like corn—and meat. Mom makes some with hot chili and some without, because our friends who come from town don't all like spicy food."

Julia put eight various sized bowls on the table and as they filled with the cut chunks of beef, Erin said, "Either everybody must take home a doggie bag or we're going to feed an army."

Julia grinned. "Lots of people visit on Feast Day."

"And most of them don't eat for a week beforehand," Neyse added.

Erin glanced out of the living room window at Tony and Carlos, who were in the yard chopping cedar wood that

Tony had gathered from the foothills.

Carlos stacked half of it by the beehive oven and the rest by the crude grate stove Julia had built by the adobe wall in the back yard.

"It's done, Mom," Carlos said, leaning his head on the screen door.

Julia went out and built a large fire under the grate. She watched it for a few minutes before coming back into the house.

"It'll be ready shortly," she said.

"We're finished with the meat," Neyse told her mother.

Julia divided the beef into several large cooking pots, put the four smaller ones on the stove, and carried the three biggest pots out to the grate.

"I'm making both red and green chili stew," she told Erin. "You can try both and see how they suit you. Some people can't take it hot but your dad, for one, loves it."

They cooked beans in one pot and boiled potatoes in another. The girls prepared the ingredients for the potato salad while Julia made prune pies, flat dishes much like pizza except that they had a top crust.

"This is as much work as getting ready for Christmas," Erin noted.

Julia nodded. "Just exactly!"

"I don't know if I can wait until tomorrow to taste it all," Erin said, catching the aroma of the chili.

"Don't worry," Julia reassured her. "We'll have to do taste tests."

Tony came in and poured himself a glass of ice water. "All the wood is done."

Julia kissed him on the cheek. "*Nachra.*"

"Carlos helped. He's gone to find Luis. Is there anything else you'd like me to do?"

"Yes," she said without hesitating. "Turn down the heat."

Erin noticed that the fire in the yard blazed fiercely,

sending heat as far as the house.

Tony looked at Julia with concern. "I would if I could. How about a glass of ice water instead?"

"That would be wonderful."

"Do you want us to check the dough?" Neyse asked her mother.

"I guess it's about time," Julia said.

Neyse pulled the blankets and the waxed paper off the tub. The dough, which had risen to almost twice its original size, smelled heavenly.

"It's ready to punch down," Neyse announced.

"It sure is," Julia agreed, looking out through the screen door.

Tony helped them carry the tub inside, where the girls had the fun of punching down the dough to remove the air. When they were done, Julia covered it again to let it rise once more.

"Better get the fire started in the oven," she said, going out the front door.

Neyse rooted in the refrigerator until she found a Coke. She put ice into two glasses and poured half the Coke in each.

"Sure is hot," she said, handing Erin a glass.

"It sure is," Erin said, watching Julia working in the yard.

After building the fire in the beehive oven, Julia again checked the dough.

"Looks like it's ready," she told them.

She and Tony carried it into the kitchen for the second time and put it on a stool in the center of the kitchen. Julia uncovered it, pulled off a large piece, and began kneading it on the breadboard using the heels of both hands. Neyse tore off a piece of the same size and after some encouragement, Erin did the same.

Julia rolled and shaped and made beautiful smooth loaves that she placed in greased pie tins. Neyse's loaves were not quite as smooth but she worked almost as fast as her mother. Erin was very slow and clumsy, but she enjoyed the effort.

"Can we try making other shapes, Mom?"

"If you'd like."

Neyse made a braided loaf and another with a clown face on it. Erin put a small swirl in the center of one and made one into a pinwheel. They put the loaves, twenty-seven in all, on a large board on the dining room table. Once again, Julia covered them to let them rise.

"Now we can light the fire."

Julia explained to Erin that the ovens in the village were actually built by two of the old ladies of the pueblo. "They use pumice stone from the area and cement them with adobe clay, which is made from the dirt. About once a year every oven needs to be replastered. I did ours last month so it would be in good shape for the feast."

Erin looked into the blazing oven. "How are you going to get the bread in there?"

Julia laughed. "The fire isn't to bake the bread: it's to heat the oven. The wood is burned down to coals and left until it starts to turn ashen. Then it's scraped out. The bread bakes from the heat retained in the stone. It'll be at least an hour before it's ready."

They went back into the house to put the fruit pies in to bake. The stews bubbled on the stove, and required occasional stirring. Julia made Jell-O for molded salads while Erin and Neyse cut fruit to put in them.

They worked without slowing, the time passing quickly.

"The coals are ready," Julia said after checking them about an hour later.

Neyse filled a bucket with water and carried it to the beehive oven. Julia took a long pole that had a heavy rag tied to the end, and dipped the rag into the water. She heaved the dripping "mop" into the oven and pulled the glowing coals out onto the ground. Neyse raked the coals to one side, and covered them with dirt as Julia pulled out more.

Once the oven floor was clean, they all helped put the loaves inside. Julia used the pole to push the pans to the farthest point inside until all the bread, as well as the prune pies, were enclosed within. Then she leaned a board against the opening, covering all but about a third of the hole.

"That's it. They'll be done in an hour."

The aroma of bread was in the air, hot, fresh, tantalizing—inside and out it was in the air, for the entire pueblo was baking bread.

"Let's have a sandwich," Julia said, wiping her face.

"And a cold drink," Neyse added.

"And sit down for a while," Erin chimed in.

Tony joined them at the table. They ate quietly and quickly, since the day's work was far from finished. Twice during the next hour Julia checked the bread. When at last it was done, she used a giant flat spatula to remove each loaf from the oven. She placed them on the same board they'd used to carry them outside, to take them back into the house. The tops were smooth and glossy, the color of coffee barely touched with cream. Even the girls' art bread looked appealing.

"Can we eat some tonight?" Neyse pleaded.

"Of course," Julia said. "Who could wait?"

They put the bread loaves and pies on a blanket on the living room floor to cool while they cleaned up the multitude of pots and pans they'd used during the day. Beans, stews, salads and posole were stored and the kitchen finally looked almost normal again.

For dinner, Julia served sliced cheese, cold meat, and giant peaches, but the real treat was the fresh, warm bread.

"Are you a certified bread maker now?" Tony asked Erin.

"Yes," Erin said with a grin. "But I can only make it in batches fit for an army."

He had not acted any differently toward her since the

night she'd asked him where he was going. She was glad, for she loved him and she would not have wanted him to be angry with her.

"What are you girls doing tonight?" he asked.

"Reading," Neyse told him. "Mom gave Erin a copy of *Laughing Boy* and I'm reading a mystery."

"*Laughing Boy?*"

"It's a wonderful love story," Erin said dreamily.

"Yes, about a Navajo. You might want to look at *The Delight Makers.* It's about our people."

"Oh, Daddy, you sound like a teacher," Neyse complained.

"Living with one does that to you," he teased.

"I don't mind," Erin said. "I'd like to learn more about the pueblo."

She cut her sentence off short, wondering if he'd taken her the wrong way. He hadn't, but she still reminded herself not to appear overly curious.

Tony and Carlos went off together that night, saying nothing to the three women when they left. Erin felt happy that she knew not only where they were going, but that Paolo would be there as well.

After texting Taylor and Heather about the day's activities, Erin read until bedtime. She imagined Paolo and herself as the main characters in the book, embracing each other with sweet tenderness. That night she dreamed of Paolo, his strong arms wrapped around her as they danced to the pueblo drums

.

She woke the next morning, a bubbling excitement in her stomach.

Feast Day.

She heard Tony and Julia talking in the kitchen and wondered how long they'd been up. She climbed out of bed, dressed quickly, and slipped out of the bedroom.

"You're up early," Julia said. "It's only seven o'clock."

"Guess I'm excited."

"It's a shame that Neyse isn't dancing," Tony said.

"It is," Julia agreed. "But, at least this way Erin has someone to watch with."

Tony gathered items from an old trunk in the corner of the living room and placed them in a heavy canvas bag. She watched him, trying not to be too obvious. Carlos added some things to his father's, the two of them talking quietly.

Erin wanted to know when the dance would begin, but she wasn't sure if she should ask, so she kept silent.

Her uncle and cousin ate breakfast, took the white bag, and left.

Within only a few minutes, a knock came at the front door. Julia answered it. Three men, dressed in traditional garb, spoke to her in low voices, asking her for bread. She gave them three of the best looking loaves.

When they left Julia told Erin, "Every household contributes food for the dancers to eat at lunch time. I was really pleased the first time they came and asked me. I felt as if they'd finally accepted me as part of the pueblo."

"When will they start dancing?"

"Not for a while," her aunt replied. "Probably around eleven."

The morning dragged along as if it had nowhere to go. Neyse was unusually quiet after she got up and Erin knew without asking it was because she couldn't dance.

The girls helped Julia put the chili, posole, and beans into pans to heat. Julia made two chocolate cream pies and a huge green salad. Neyse filled three large jars with water, hung tea bags in each, and set them on the patio outside to steep. Erin made Kool-Aid while Julia sliced a watermelon.

All was prepared, but no one arrived.

"I think I can hear the drums," Neyse said after a time. She went outside and came back in, excited. "They're starting. Let's go!"

Erin jumped up but Julia didn't move. "You two go

ahead. I'll wait here for a while."

She left her cell in the bedroom. The dance was no place for a phone.

As they walked down the road toward the church, Erin was aware of how hot it was already, and was glad that she had put her hair in a ponytail as Neyse had suggested. She'd tied a bright green ribbon in it that matched the apple green sundress she wore. Neyse had on a calico skirt and a tomato red blouse. She'd braided her hair.

The beat of the drums grew louder and louder as they drew closer to the church. Men's voices chanted in singsong syllables to the rhythm of the drums. Erin felt as if the drums were beating inside her chest and found herself, like her cousin, walking with the beat.

As they passed along the side of the church Erin caught her first glimpse of the dancers, their richly clothed bodies moving as one. There were two long lines of men and women, alternately, the tallest in the center, the children on either end. She scanned the faces looking for Paolo, but she didn't recognize anyone.

"I don't see Paolo," she whispered to Neyse.

"That's because this isn't our side. The pueblo is divided into two clans. We're the Pumpkin side. Turquoise side dances first, then the Pumpkin side."

They sat on the low adobe wall in front of the churchyard and watched the dance. At last Erin saw what she sensed was truly Indian, what she hadn't seen at Taos, what was private and sacred. The dancers' faces showed intense concentration, with no recognition of anyone in the large crowd who watched. Erin felt as if she were a participant in an ancient ritual few outsiders ever observe.

The men and boys were clothed similarly: heavy white woven cotton kilts trimmed in black, green, and red. Around their waists they wore leather belts on which were fastened bells in various sizes. From the belt in the back hung the skin of what Erin thought must be a fox. Bare arms and torsos were painted with a deep red-brown paint, just a shade darker than their skin.

All the men wore white moccasins trimmed with some sort of fur. Just below their knees hung tassels of black and green yarn. Each man wore a clump of green and yellow feathers in his hair. Erin noticed a wide variety of neck and arm decorations, including shells, beads, turquoise, silver, and leather. Hung across from shoulder to waist were strung shells. In every case, the men carried gourd rattles that they shook in time with the drums.

The women and girls were also dressed alike, in black dresses, cut diagonally across both front and back to expose one shoulder. Bright calico undergarments showed at the hem of each dress, a red sash cinched the waist, and in most cases, they wore turquoise wristbands. Each woman was fitted with a headdress that appeared to be made from very thin wood, painted turquoise, with red, white, and yellow designs. They all wore moccasins the color of adobe dirt, and carried fir branches in each hand that they alternately raised and lowered to the drummers' beat. Erin noticed that even the youngest dancers kept time and followed the dances conscientiously.

The dance came to an end. Erin was so delighted that she instinctively raised her hands to applaud but Neyse grabbed them.

"Don't clap, remember?" Neyse cautioned and Erin nodded, embarrassed.

"I'm so sorry. I forgot. It was just so wonderful."

"It's okay. No harm done."

The first group left and the sound of another group could be heard approaching in the distance. The singers and drummers came first, followed by the dancers. Erin strained to make out the features of the men, trying to find Paolo, Juan, Tony, or Ramon. She had trouble recognizing any of them with the paint and ceremonial dress.

Neyse nudged her in the ribs and nodded toward the

dancers who were now in front of them. At that moment she recognized Paolo. He looked totally absorbed, almost as if he was in a trance. She wanted him to know that she was there, but she knew that she could neither speak nor move, for he would be angry if she intruded on the ceremony. She watched with pride as he held his head high and danced with a light step, as if anticipating the beat.

The dancers snaked around in a circle, changing directions once and then again, the drums sounding a steady cadence. How proud they looked, almost regal, yet at the same time, simple and childlike. *For how many hundreds of years, have they danced this exact same dance?* She followed Paolo with her eyes, catching glimpses also of Tony and Ramon, Luis, Carlos, and Alonzo. She found Felicia, along with the other children, dressed just like the women, looking very much like a costumed doll.

She watched them form two long lines facing each other. They danced toward the center, passed each other and continued to the opposite point. Juan and Paolo danced among the young men and women, smooth-faced youth, full of energy. Erin saw Natoma, even more attractive in her traditional dress, and she felt jealousy tugging within, for Natoma shared with Paolo a heritage of which Erin could never be a part. She pushed the feeling aside, not wanting it to spoil the day.

The drums stopped, the dancers turned and slowly left the area. Erin was keenly disappointed.

"Is that all?"

Neyse laughed. "Hardly. This goes on all day. They're just moving to the plaza."

"Can we watch there, too?"

"Sure. We'll go to the Trujillos'. They live right on the square. We couldn't get a better seat. But first I want to go home and see if anyone has come yet. Since I'm always dancing on feast days, I don't get much of a chance to visit with anyone."

They walked along the dusty road toward home, past a line of concession stands that had been brought in for the day. A small carnival, complete with mechanical rides and games, was operating not far from the plaza. Erin felt the sickness she'd had when she'd visited Taos, of the outside intruding on the pueblo.

Why must they spoil it?

Why did they come here?

Balloons, shirts, cotton candy, cheap toys, trappings of the outside world. It had nothing to do with the feast. It didn't belong here.

She was surprised to realize how much it annoyed her.

When they arrived back at home, the house was crowded with people. Neyse introduced Erin to two families from a pueblo to the north, distant relatives who always came for the feast. The guests ate together at the table while Julia hurried back and forth, filling bowls, cutting bread, and pouring drinks. The girls washed the dishes after one group of guests left, just in time to greet several more visitors; outsiders and friends Julia knew from the university.

Neyse cut Erin and herself big squares of prune pie. It reminded Erin a little of a packaged breakfast snack except that the sweet filling and flaky crust tasted far better. She snuck another slice and Neyse laughed at her.

"Your stomach is turning Indian."

After the next batch of dishes, Julia urged them to go and watch the dancing. "You can do dishes every other day. Go and watch the dance while you can."

They sat under a thatched roof in front of the Trujillos' house, along with about a dozen others, glad to have some relief from the blazing sun. Their side danced again, the same two patterns of circles and straight lines with subtle embellishments that Erin began to detect.

An old man, with more lines on his face than a city map, sat silently in a folding metal chair. Neyse told Erin that he was Mrs. Trujillo's grandfather.

The family didn't know how old he was, but they thought he was at least ninety-five. Although most of his teeth were missing, the delight on his face when he smiled made everyone else around him smile, too.

"Did you see that fancy red car parked down near the church, Grandfather?" Mrs. Trujillo asked. Her arthritis kept her from participating in the dances.

It was a while before he answered. "I saw it."

"That was a Mercedes, Grandfather. Antonio told me it cost $100,000."

The old man looked at her with disbelief verging on disgust. He considered what she'd told him. Then he said, "It'll only rust."

Mrs. Trujillo tried to keep a straight face as she nodded in agreement. The rest of the group laughed loudly.

Erin knew that they were laughing, in part, at the white man's values, but the honesty of his remark took the sting out of it. She laughed with them.

The smallest children were beginning to tire, leaving the line of dancers to find a lap somewhere among the onlookers. Felicia looked weary but kept up as best she could with the others. Erin spotted Victoria, beautiful and serene, seeming not the least worn out.

The dances lasted about half an hour, and then there was an equal time of rest while the other side danced. Had it not been for the heat, dancing for that long might have seemed tolerable. But Erin knew it was well over 100 degrees and even in the shade it was nearly unbearable. She wondered how they kept up the pace.

During the mid-afternoon break, Tony, Carlos, and Juan went back to the house to eat and rest. Neyse and Erin again helped Julia serve. The men sat at the table together, talking quietly with each other about the dancing. Erin overheard their critique of both *Kivas'* dancing and was amazed to realize she'd missed so much.

Erin hoped that Paolo might come as well, but the men

left shortly after eating and he had not appeared.

A never-ending stream of people, Indian and Anglo, came and went, eating, talking, laughing, and sharing both food and friendship. Julia never seemed to stop and Erin began to feel sorry for her, for she looked exhausted.

"Are you ready to go back to the plaza?" Neyse asked Erin.

"Uh..." She tried to think. "I have a headache...not bad or anything, but I think maybe I'll rest for a while."

"I can stay here with..."

"That'd be silly. Go on back to Trujillos' and I'll come in a while," Erin insisted.

"Okay, if you're sure."

"I'm sure."

"Okay then, I'll see you later," Neyse said hesitantly, standing at the door.

After Neyse left, Julia looked at Erin with surprise. "Aren't you going with Neyse?"

She hesitated, knowing Julia would not want Erin to stay home for her benefit. "I don't feel too well."

"Maybe it's the heat," Julia suggested. "I can't be out in it too long or I get sick, too."

The last group of friends had just left and Julia was relaxing on the couch before cleaning up. Erin sat down next to her.

"Are you going to go watch the dancing?" Erin asked.

"I am, but not until it cools down some. What did you think of it?"

"I loved it," Erin said. "I felt like I was right with them; like I was a part of it."

"It's very special."

Julia was quiet for a while, looking out the window. "I like having everyone here for feast. That's a lot of fun. But I prefer the closed dances to the open ones."

Erin nodded. "I think I understand why."

"My very favorites are the Christmas dances, which are

presented on the four days after Christmas. Generally they do animal dances—deer, buffalo, or antelope. They use a lot of fresh greenery with their costumes. When it snows, the sound of the drums is muffled and everything and everyone is dusted with white. It's like a winter fantasy."

Erin tried to imagine snow on the pueblo, but the heat made it impossible.

"I'll never forget the first time Carlos danced," Julia said. "He was only about eight at the time and he was so young...to me, at least. I went to watch and took him a small bag of food. I gave it to him but he didn't say anything. Later, when he told me about his day, he said that a nice lady had brought him some food. He hadn't even recognized me! I guess that the dancers are almost in a sort of trance, not altogether aware of what's going on around them."

"Did you tell him that you were the lady?"

Julia shook her head. "No. There was something magical for him that a stranger cared that much for him. I just let it go."

"I wish I could have seen him," Erin said wistfully.

"I have a picture of him, somewhere," Julia said. "Let me look. I think it's in the bedroom."

Erin followed her down the hall and sat on the bed while Julia looked through a box from a shelf in the closet. The woman in the photograph that she'd noticed the day she'd arrived, stared at her.

"Who is the lady in the picture?" she asked Julia.

Julia looked around. "She was my grandfather's sister, a relative of yours, too. I never met her. She died when she was only nineteen, but I always loved that picture. I don't know much about her except that the family refused to allow her to marry the man she loved. They thought his background was beneath them. Grandfather used to say that she died of a broken heart."

"I think you look like her. She's really pretty."

Julia smiled. "Thank you."

She found the picture and handed it to Erin. In it, Carlos stood in front of the house, fully costumed.

"He does look young," Erin agreed.

They looked at it for a few moments and then Julia put it away again.

"Why don't you rest a while," Erin suggested, "and I'll go do the dishes."

Julia looked hard at her. "Is that why you stayed home?"

"Uh..."

"Don't be silly. I do this all the time. You go along and watch the dances." She steered Erin out to the front door. "I'll come along in a little while."

But she never did. Erin didn't know if it was by chance or by design...whether she preferred being at home with friends rather than at the dance with outsiders. Perhaps she felt too much like an outsider herself, watching her family dancing a dance she could never join.

CHAPTER 11

The sky grew dark, clouds piling up one on top of another. Lightning flashed streaking bolts in the mountains far to the west.

"Do you think it will rain?" Erin asked.

Jeanette, the daughter of friends living in a nearby pueblo, said, "It always rains here after they do the corn dance."

Erin and Jeanette sat with Neyse on the side of a pick-up truck in front of the house. The dance was over and everyone but Jeanette's family had already left. The younger children had run off to the carnival, hardly able to contain their excitement at having somewhere to spend their money. Jeanette's parents were inside talking with Tony and Julia.

"Is Juan coming over?" Jeanette asked Neyse.

"I hope so. It depends on how tired he is, I suppose."

That was when Erin saw the figure coming down the road. Even if she hadn't seen Tiya, she would have recognized the gait. Neyse saw him too.

"Well, we know that some people aren't too tired."

He wore jeans once again, with a pale blue plaid shirt, and though she knew it was silly, Erin was disappointed. She had hoped to see him up close, dressed in his native costume. He came through the gate and up the driveway, smiling almost with embarrassment at meeting all three girls.

They exchanged greetings.

"I thought Juan would be here," he said to Neyse.

"I haven't given up hope yet."

He was quiet for a few moments.

"Would anyone like to go for a walk?" He said it to them all, yet it was clear that he was speaking to Erin.

Jeanette spoke quickly. "We're probably going to leave pretty soon."

Neyse shook her head. "I think I've done enough walking for one day," she said, wiggling her foot. "Why don't you and Erin go?"

Erin wanted to do just that but she didn't want to be rude to Jeanette and Neyse. "Are you sure?"

"Of course. Go ahead. I'll tell Mom and Dad."

The evening was at the halfway point, being neither dark nor light, requiring the eyes to work twice as hard to see even half as much. It was still warm, but Erin was glad that she'd changed into jeans and a long sleeved tee shirt, for the wind had a distinct chill. They headed toward the canal, though not by way of the carnival. Firecrackers exploded with regularity and Tiya barked each time. The wind had cooled the air, reminding Erin of evenings in California.

When they reached the canal Paolo bent toward the dog. "Go home, Tiya." The animal looked down, obviously not happy. "Go."

Tiya turned and walked slowly down the road.

"You were at the dance," he said, not really questioning, yet asking for a response.

"Yes. I saw you."

"I felt you there."

"It was wonderful, but it looked like hard work. Are you very tired?"

"A little."

Erin felt strained, as if something were missing or going unsaid, but she couldn't put her finger on what it was. They crossed the canal and walked down the path by the fields.

"Did you go to the carnival?" he asked.

"No."

"Why not?"

She looked at him. "I didn't want to. It doesn't belong here. It's like bringing a circus to church."

He didn't say anything, but he put his arm around her waist and pulled her closer to him. Her heart tripped over itself.

He led her farther away from the village, across the fields to a grove of cottonwood trees. He found a clear spot by a tree and sat, gesturing for her to do likewise. The river sang to them, drowning out the sounds of the village.

"You're different," he said after a time.

"Different from what?"

"Different from other outsiders."

"How?"

"You think like I do."

"That's pretty different," she said.

He tensed.

"I'm only teasing you," she added gently.

"Lately, it seems like you understand me without my having to explain things."

"Sometimes I do. But other times I don't get you at all."

"When?" he asked, sounding wounded.

"Sometimes when you get mad, I can't figure out what you're so upset about."

He paused and then admitted, "I don't always know myself. Sometimes I react to things without giving it enough thought and my mouth gets carried away. When I think about it later, even I can't always figure out what set me off." He looked at her. "Tell me about when I got mad that you couldn't understand."

Erin hesitated, not sure she wanted to open a can of

worms and spoil this beautiful night. "Well," she began, "whenever my phone pings you have a fit."

"Ah," he said, "that."

"I mean, I never stay on it for long. Sometimes I don't even answer."

Paolo settled himself. "It's the presence of it…the need to be connected elsewhere, as though being exactly where you are isn't enough; as if the company of the people you're with is lacking and you need something else or someone else. It's not like there's an emergency or a crisis. You just need something…more. It's worse than rude—it's insulting."

He said it all in a quiet voice, not the nasty one she'd grown to dread so much.

"It isn't just when yours rings. It's everyone's. Outsiders are so connected to their electronics that they lose sight of reality. They take computers or iPods or iPads or cell phones everywhere they go. They think they can't be unconnected. They think they can text and drive, for heaven's sake. How many have died because of that? They can't seem to walk down the street without being connected to that giant transmitter in the sky. It's nuts and when I see it here, it makes me crazy."

Erin was utterly silent. She wasn't sure if she should try to defend herself or to simply say she understood.

"Don't be upset," he pleaded. "It's not you—it's me. I just see things differently. Most of the kids here have cell phones. It's the way of the future. I just think they ignore our key values when they use them like outsiders do."

"Like I do," she said softly.

Reluctantly he nodded. "You don't like it, but my world and your world are different."

After thinking a while Erin said, "When you explain it like that, I can honestly see what you mean. I'm not just saying that," she insisted. "It actually does seem sort of rude. I'm just so used to doing it because everyone else does." She smiled wistfully. “My mom always says, ‘If everyone else jumped off the Golden Gate Bridge, would you jump, too?’”

“Good question. Your mom sounds pretty smart.”

“She’s not bad, for a mother.”

He smiled at her. “What else does your mother always say?”

Erin thought a minute. “She always says, ‘Floss your teeth every night; you only get one set of teeth.’”

“Very profound,” he said, nodding in agreement. He reached around and tickled her in both sides. She jumped and squealed. “Ah, so you’re ticklish.”

She retaliated but with no success.

“Indians aren’t ticklish.”

She looked hard at him. “Really?”

“Really.”

She wasn’t convinced. They wrestled playfully with each other but his strength far exceeded hers and he easily pinned her down.

“Give?”

“For now.”

“I could keep you here for days, you know.”

“I doubt it.”

“You do?”

“Yes. Antonio would come get me.”

“He wouldn’t even notice you were gone.”

“He would too. I’m his favorite niece.”

“Wait till I tell Felicia what you said,” he teased, starting to get up.

“Oh, no, Paolo, don’t you dare!”

He sat down. “I’d never hurt Felicia,” he said quietly.

“You really love her, don’t you?”

He nodded.

“It shows.”

“Do you have a boyfriend?” he asked suddenly, changing the subject.

The question completely surprised her. It didn’t at all

seem like the sort of thing he'd ask. She shook her head slightly. "No. Do you? Have a girlfriend, I mean?"

"No."

The long silence that followed made her feel uncomfortable and sorry she'd asked him.

He grinned. "I wonder what's the matter with us."

Erin began to giggle. "I *know* what's the matter with you. I, by comparison, am perfect."

"You're too skinny to be perfect."

"Well," she harrumphed, and started to get up.

He grabbed her, pulling her down beside him. He loosened his hold, leaned over, and kissed her full on the lips. Shyly, she put her arms around him, feeling the broadness of his back.

He kissed her again and again, searching her face with his lips, asking for her kisses in return.

She was uncertain. She'd never before been kissed that way...never been willing to let anyone kiss her more than once or twice, in fact. This was different and it was scary. She felt a strange trembling in her stomach and a wild desire to push hard against him.

"Erin, you're really special," he whispered.

"You are, too."

"I like you. I like being with you."

"I'm glad. I didn't tell you before, but I liked watching you dance."

She said it without considering his response and didn't realize how touchy the subject might be until it was too late.

He pulled back slightly from her and looked at her uncertainly. "Why?"

She thought a moment. He didn't sound mad and she certainly didn't want to make him mad. "I'm not sure," she said at last.

"Did you like watching Uncle Tony?"

"Yes..."

"Who else?"

"Juan and Carlos and your brothers. Everybody. I liked watching everyone. Even the people I don't know. But I liked watching you best. You were so intense—so completely in that moment; it was beautiful and I loved it."

His eyes smiled at her. He said nothing. He kissed her again.

They lay on a cushion of leaves beneath the trees, oblivious to all but each other. Erin yielded to his kisses, melting into her excitement. She felt dizzy, as if she weren't quite connected to herself.

He kissed her cheeks and her neck. He whispered her name.

He slipped his hand up, slowly, along her side, stopping when he reached her bra. Slowly he moved his hand over until he cupped her breast.

Instinctively she moved, shoving his hand aside. Her mind jolted back to reality. She jerked away from him and sat up.

He was beside her instantly. "It's okay," he soothed her.

"No."

"I won't do anything you don't want me to."

She pulled her knees up against her chest, wrapped her arms around her legs and rested her chin on her knees. She wasn't sure what she wanted.

"Erin," he whispered. "I won't hurt you."

She wondered how it would be to let go, but she knew she would not; she knew that letting go might very well lead to trouble.

Paolo put his arm around her waist, uncertainly. "I'm sorry, Erin. I guess I got carried away."

She didn't respond.

"I wish you'd say something."

She couldn't. She was too confused. Her feelings didn't make sense.

He sat beside her, waiting. "Do you want to go back?" he asked at last.

She nodded.

Paolo stood and helped her to her feet. He put his arms around her. “Please don’t be upset.”

She put her head against his chest and he held her, bending his head down to her shoulder and burying his face in her hair. She put her arms around his neck, returning his embrace. Tenderly he took her by the hand and led her back toward the village.

They walked slowly, feeling the cool calmness of the night. A dark, cloudy sky hid the moon. Only a few lights were still burning, most houses already dark and quiet. They walked without talking; Erin felt sad and uncertain.

A myriad of tiny bugs flitted around the porch light at Neyse’s. Just before they reached the porch, Paolo pulled her to a stop. He put his arms around her, started to say something and then stopped.

“Are you mad?” he finally asked.

She didn’t answer right away, and at last said, “No.”

“Are you afraid of me?”

She hesitated. “A little.”

Paolo took hold of her shoulders and looked hard into her face. “I won’t push you again, I promise. I don’t want you to be afraid of me.”

She wanted so much to believe him.

“Will you see me again?” he asked.

She looked down and then back up at him. “Yes.”

He kissed her gently. “I’ll come by tomorrow then, all right?”

“All right.”

He watched her as she opened the front door and went into the house.

To her great surprise, Julia sat on the couch, reading.

She looked up. “It’s pretty late, Erin.”

Erin swallowed. “I’m sorry,” she said, hoping Julia wouldn’t get mad at her.

“I was concerned about you. I don’t want anything to

happen to you." She patted a place beside her on the couch. She was quiet for a time and then she said, "Things are different here than they are other places. People think differently. You already know that. Sometimes, things move faster than young people are ready for them to go. Do you understand what I mean?"

I sure do, she thought. "Yes," she answered without comment.

"I have a responsibility to your parents as well as to you while you're here. I don't want to let any of you down. You have a responsibility to yourself, too, Erin. Even small decisions can make a big difference in life. Don't ever forget that." She smiled and patted Erin's hand. "I think it's time for bed."

"I'm sorry if I kept you up."

"Don't worry about that. I wanted to finish this book anyway. Good night, dear," she said, and gave Erin a loving hug.

Erin tiptoed into the bedroom. Neyse was already asleep. Erin stripped down to her underpants, put on her nightshirt, and slipped under the covers. She turned onto her stomach and wrapped her arms around her pillow.

How wonderful it had felt to kiss him...how exciting to have him hold her. She could have gone on kissing him all night long. *Why did he have to touch her like that and spoil it?* Would she end up fighting him off like she had that creepy Jeff Anderson back home?

Then the part of her that had wanted to let him continue pressed into her mind. What would it be like? How would it feel, to be touched that way? *How does it all happen?* She'd always known it couldn't be like the lame movies they showed in biology. And it certainly wasn't like the dogs she'd seen down the street. It *must* be different than that, or people wouldn't like it so much.

She could just imagine her mother's disapproval, her father's anger, and Scott's disappointment. She thought about her friend Karla, whose sister had gotten pregnant when she was a senior. The entire household was in an uproar for months; it seemed like Karla, her mother, and her sister cried the entire time. They had wanted to keep it a secret, but practically everybody knew by the time she went away. It had been awful.

One thing for sure, she thought. *I don't* ever *want to get into a mess like that.*

She was still thinking about it when a soft rain whispered down the window and she fell asleep.

CHAPTER 12

"It isn't much farther now," Juan said, helping Neyse as she slowed down, favoring her weaker ankle. He was taking the girls to his favorite fishing spot.

"Exactly what does that mean," Neyse asked, "in terms of hours?"

Erin giggled.

"Oh, come on, *Saucha*, five minutes, max," Juan said.

Neyse looked at her watch and said, "I'll keep track."

"What does *saucha* mean?" Erin asked, hopping from a boulder to a large rock and onto another boulder.

"*Saucha* means child," Juan told her, patting Neyse on the top of her head. "Sometimes we use it as a nickname."

"Keep it up," she threatened, "and you'll swim the rest of the way."

The late afternoon sun peeked in and out from behind cottonwood trees as they walked along the river. The glossy-green, heart-shaped leaves glimmered in the flickering light.

"I'm not that crazy about fish in the first place," Neyse said, slowing her pace even more.

"We'll wrap them in bacon and broil them. You'll love 'em," Juan assured her.

"Nobody 'loves' fish. They eat them if forced, at best they like them. But absolutely nobody 'loves' them."

"I like fish," Erin argued.

"See? That's exactly what I mean. Erin said 'like' not 'love.' You don't 'love' something that might choke you to death."

"Neyse," Juan said, sounding exasperated, "we'll put yours in the blender. You'll never even know it had any bones."

"Yuk!"

"Even steak has bones," Juan argued.

"When was the last time you got one caught in your throat?"

Erin laughed and nearly lost her balance.

"I give up. We'll make you a cheese sandwich."

"Yuk," Erin said.

Neyse laughed.

"You two are a real matched pair," Juan commented. "In any case, here we are," he said with a sweeping arm gesture.

"This is it?" Neyse said. "It doesn't look any different than a hundred other places along the river."

"Ah, but it is," he assured her.

Juan took off the backpack he was carrying and opened it. He pulled out and put together three collapsible fishing poles, handing one to Neyse and one to Erin. He opened a container, alive with fat red worms.

"Now," he began, "this isn't as bad as it looks."

"We've both fished before," Neyse told him impatiently.

"*Hin'a*," he said.

Neyse translated the new word. "That means okay," she explained.

The girls baited their hooks and dropped their lines into the water.

After about a minute and a half, Neyse said, "Well,

where are all the fish?"

Juan shook his head. "Be patient, Neys."

"I have been patient."

"Did you expect them to jump into your lap?"

Erin and Neyse both laughed.

"Well, the way you talked about it, this is the premiere fishing spot of the country."

"One reason I've always enjoyed it so much is that it's quiet here," he said with a meaningful smile.

Neyse stuck her tongue out at him, then turned her attention to the river.

Juan caught two fish in rapid succession and within five minutes Neyse got a bite on her line.

"Yank it up," Juan whispered loudly, "and get a good hold on it."

She gave him a quick sideways glance. "I can manage, thanks." In less than a minute, she'd pulled in a trout about ten inches long.

"Not bad," Juan said approvingly.

"Not bad?" She pretended to throw the fish at him.

"What's holding you up, Erin?" Juan teased.

"Nothing. Out in California we don't much bother with minnows like these. We throw anything under three feet back and wait for a real fish."

A familiar voice came from the hill behind them. "Last month she caught a killer whale."

They all turned to see Paolo sitting high on the bank above them.

"How did you ever find us?" Neyse asked with surprise.

"Oh, I've known about Juan's favorite fishing place for years," he said, making his way down to them.

"You certainly got to it the hard way," Juan observed.

"Not really." Paolo sat down near Erin. "I didn't want to walk along the water and spoil your luck."

"Where's Tiya?" Erin asked.

"Fishing is one of the places Tiya doesn't get to go."

"We took him once," Juan said, "but we couldn't keep him out of the water."

"Tends to distract the fish," Paolo added.

The two friends laughed, remembering.

"Did you bring a pole?" Juan asked.

"No, I only came to watch."

"You can use mine," Erin offered.

"Are you kidding? I expect to learn a lot just watching you. It's a rare opportunity to see a Californian fish and I wouldn't think of missing it."

She felt self-conscious with him sitting right next to her. He didn't talk at all for a long time and that made it worse.

Erin felt a tug on her line. She waited until she felt it again and then pulled quickly on the pole. The fish pulled back.

"Got one?" Juan questioned.

"I think so," she said.

She gave it some line, then started reeling it in, slowly and steadily, hoping she wouldn't lose it.

The fish flipped wildly when she pulled it from the water, but she grabbed it and held it tightly while she extracted the hook from its mouth.

"It's a beauty," Juan exclaimed. "Trout, I'd say twelve inches."

Erin handed the pole to Paolo. "Your turn."

"No, you go ahead: you're pretty good at this."

She shrugged. "*Hin'a*," she said, pleased she'd remembered the word for okay. She baited the line again and tossed it into the water.

In all, they caught seven fish. They stayed until the sun set and then walked back to the road where Juan had parked his father's truck.

"How did you get here?" Erin asked Paolo, expecting to see another car.

"I started walking," he explained, "but Mr. Ramirez came by and gave me a ride."

They all piled into the cab and Juan drove back to Neyse's. Paolo had his arm around Erin's shoulder on the way, but she sat stiffly next to him.

When they arrived home, Paolo said, "I'll teach Erin how to clean fish."

Juan looked surprised but shrugged and replied, "Sure, okay. We can get everything else ready."

Paolo and Erin went around to the back porch while Neyse and Juan went inside. Neyse flipped the outside light on for them and Paolo began to gut the trout. Erin stood and watched.

He spoke to her without looking up. "Are you mad at me?"

She considered the question. *Mad* didn't describe her feelings. She felt uncertain. The uncertainty frightened her. She felt scared of him; scared of liking him too much.

He finally looked at her. "Well?"

"I'm not mad."

"Something is wrong. What is it?"

She used Julia's words. "I just don't want to get hurt." After it was out of her mouth it sounded like a dumb line from a movie.

"I won't hurt you. I told you that last night and I meant it. I shouldn't have pushed you. I'm sorry. Maybe that doesn't help, but I mean it. It won't happen again." He looked at her, his eyes dark and his jaw set.

She wasn't sure where it would lead, but she knew that she liked him too much to keep the distance between them.

She smiled shyly. "Okay."

The door opened and Juan stuck his head out. "Aren't you done yet?"

Paolo shook his head. "Not quite yet, I'm afraid. Californians learn slowly."

"What?" Erin squeaked.

Paolo grinned.

"Cancel the lesson. Just finish the fish. We're already in

full production in here." The door closed.

Paolo finished the rest of the fish and together they took the prized catch into the house.

Tony came into the kitchen. "So this is the long awaited dinner."

"It's not ready yet, dear," Julia cautioned him. "Give us a few more minutes."

"Where's Carlos?" Neyse asked her mother as she sliced the bread.

"He got tired of waiting for dinner. He said he was going to see Luis."

Neyse laughed. "He was probably going to see what Victoria was cooking tonight."

"Where is that bacon?" Juan asked, moving things around in the refrigerator. "Didn't you say you'd get some?"

"It's there. Anytime somebody promises me a fresh fish dinner for the price of a pound of bacon, I'll bring home the bacon," she said, with an impish grin.

The four young people laughed.

Tony moaned.

Juan prepared the fish and put them in the oven. Together they set the table, finished making a salad and warmed the beans.

"I think we're all set," Juan said as he put the steaming platter of fish in the center of the table.

Everyone sat down hurriedly and began to dish up.

Erin felt more relaxed, now, being near Paolo. He sat next to her, touching her leg inconspicuously with his own.

A studied silence ensued as each person took a bite of the fish.

"This is delicious," Julia exclaimed.

Tony agreed.

"The bacon is good," Neyse said, smiling at Juan.

"I think it's very nice," Paolo commented. "But I must say, it doesn't compare with whale."

"Whale?" Tony repeated, but Erin laughed, happy to

have Paolo teasing her once again.

"It's a long story," Juan explained.

"Too long," Neyse told him.

"Are you going to the juniors' playoff game tomorrow night?" Juan asked, directing the question at Julia.

"I hadn't thought much about it. We've only gone to two games this summer, since neither Carlos nor Neyse are playing."

"I heard it's supposed to be a good match," Tony commented.

"Should be. Both teams have had a strong season. What do you say, Neys? Do you want to go?"

Neyse nodded and turned to Erin. "You, too?"

Erin agreed as well.

"Am I included in this or is it a private party?" Paolo asked.

Juan looked amused. "I thought you didn't like baseball."

"I don't. But I like it better than being left out."

"In left field?" Erin finished for him.

Everyone laughed.

He gave her a disgruntled look, but it was edged with humor, a look which told her they'd progressed a long way since the day she arrived.

"Well, then," Juan pursued, "are we going or not?"

"We're going, we're going," Neyse said impatiently.

"What time?"

Neyse sighed. "Juan, *ma'a!* The night games are always at seven o'clock."

"We'd better get there early if we want a good seat."

"Why don't you get there early and save a seat for the rest of us? We'll meet you there," Neyse said with annoyance.

Erin looked at Neyse. She'd never before heard her cousin speak in that tone. She glanced quickly at Paolo but he seemed not to have noticed. Juan, too, appeared unaffected but Erin thought he backed off some.

"Okay," he told them, "we'll see you there."

Juan and Paolo left shortly after dinner. Neyse immediately went into the bedroom and got ready for bed. She stood in front of the mirror, brushing her hair vigorously, tearing the brush through it with forceful strokes.

Erin watched her uncertainly and finally said, "Neys, what's the matter?"

She ripped the brush through her hair again before answering.

"Sometimes he gets on my nerves."

"What did he do?"

"He was pushy. I don't like it when he acts like that. It's as if he's an outsider instead of one of us."

Erin thought about that for a while. Neyse drew a line just as Paolo had, Indian on one side and outsiders on the other. She wondered where she belonged...where they saw her.

Neyse continued. "One of the best things about living here is the pace. It's slow, no pressure. Juan has changed in the past year, ever since he started at the university.

"He talks about going into politics. He says that's the only way to change the government's treatment of Native Americans. Once he told me he wants to be the governor of the state!"

Erin listened, at the same time thinking that what Neyse was saying explained a lot of questions she'd had about Juan, like his confidence and leadership.

"I suppose part of it is because… oh, never mind."

"Because what?"

Neyse looked as if she wasn't sure she should complete her thought.

"Juan's father is an alcoholic. I think that Juan is afraid of ending up like him. He wants so badly to do well that he pushes harder than most people here do."

Erin was learning that by listening to the spoken as well

as the unspoken language, she was understanding this culture better every day.

"Most of the time it doesn't bother me, but now and then I feel like slugging him. I don't want to be tense and tied down to deadlines."

"But without plans, people wouldn't ever accomplish anything."

"That's dumb. Of course they would. People decide what they want to do and they do it. I'll go to the game, but I don't want to have to be there at a certain time. What difference does it make what time we get there?"

"What about a movie? Would you want to be on time and see the beginning?"

"Probably. And I'd want to be at the airport when my plane left, if that's your next question."

Erin grinned. "It was."

"I know that schedules are necessary, but I want to make sure the outside world doesn't control my life. I don't want my spirit killed by deadlines."

"I have a lot of deadlines at home," Erin said. "School, swim team practice, meets, piano lessons, football games, even parties and dates. Just about everything I do. I don't think it kills my spirit. Most of those things are really exciting to me."

"If you're not careful, you'll turn into a robot. At eight every day you'll be at school. At four you'll be at swim practice. At seven every night you'll practice piano. At ten every Saturday you'll be at a swim meet. At three every Saturday afternoon you'll be at your piano lesson. Ten years of that and you could do it in your sleep. That's dead! No life left in it."

"It sounds terrible when you put it that way," Erin agreed.

"Deadlines kill people. It's a fact."

"Still, I wouldn't be a very good swimmer if I only practiced every now and then. I certainly wouldn't have been on the swim team."

"You're right," Neyse said, sounding irritated. "Let's forget it. I'm tired." She got into bed and turned toward the wall.

Unhappily Erin switched off the light and climbed up to the top bunk. She slipped under the covers, taking care not to shake the bed more than necessary. She wanted to understand what Neyse had been trying to say. It seemed almost as if they'd been talking about two different things. Some deadlines were necessary. Neyse agreed to that. Did she mean that too many deadlines were bad? Was that it? Or maybe it was that if they weren't necessary, you shouldn't be controlled by them, like the baseball game. But the players had to arrive at the same time. You couldn't start a game with only three people. Where did you draw the line?

She frowned to herself, still uncertain.

Sometimes when she was alone, she found herself thinking about philosophical things…things like life and death and the universe...about what really mattered. But she didn't have many friends who were interested in those elusive things. At least they didn't tell her about it if they were. She was glad to have had a chance to talk about them, even if it hadn't turned out so well.

The next morning Neyse greeted her cheerily and acted as if their disagreement had never happened. Erin was relieved to be on good terms again, but she felt disappointed that they'd never resolved their differences. *Perhaps some things you just don't agree on.*

The baseball game started late that night after all and although Erin and Neyse didn't arrive at the field until well after seven-thirty, they still saw the first pitch. Neyse gave Erin a satisfied grin when they discovered that the game had been delayed, but neither Juan nor anyone else in the crowd seemed the least bothered at having waited for almost an hour.

"My mom always says, 'You can't hurry life, it'll happen when it's ready,'" Erin whispered to Neyse as they climbed up the rough wooden bleachers to where Juan and Paolo had saved them a seat.

"Did you have trouble finding the place?" Juan teased.

Neyse hit him on the arm.

"Actually," Erin said, sitting down next to Paolo, "we've been stuck in traffic."

Paolo gave her a smile that said, *Nice comeback.*

A woman sitting behind them spoke to the man next to her. "I hear there's a scout from the major leagues here tonight."

Juan chortled. "He must be here on vacation."

Paolo laughed and pointed at Tiya, who ran happily back and forth along the first base line until the umpire yelled at him.

He slunk back to the bleachers and sat.

The game was close, one exciting play following another. One young boy, a master at stealing bases, kept the crowd in a frenzy. By the time it was over, Erin's voice was nearly gone. The fact that their team lost hardly made a difference.

As they walked home together, Erin noticed that Juan held Neyse firmly around the waist, leaning over frequently to whisper in her ear. It looked as if the disagreement of the previous evening had been totally forgotten.

"What are you thinking about?" Paolo asked, pulling her to a stop.

"Oh, nothing important," she said. She looked up at the quickly darkening sky. "It's a pretty night, isn't it? It looks like God sprinkled stars in the heaven."

"That sounds almost like a poem."

They resumed walking.

"Like Robert Frost, I hope."

"Oh? Why Robert Frost?"

"I did a term paper on him in my Literature class last

year. I love his poetry. Do you know any of it?"

"Only one or two."

"Which?"

Paolo paused for several moments and she was afraid that he didn't know Frost at all...that she'd put him on the spot.

But then he said, "The one about a road not taken is the one I like best. Life is always one choice after another, especially for Indians. If we take the outsiders' way, we have to turn our backs on our own."

"We all have to make choices, Paolo."

"It's different for us."

"Maybe," she said, again not sure that she agreed.

By this time, they'd reached Neyse's house and Paolo told Tiya to stay outside. The dog sat obediently. Waiting inside on the dining room table was a pineapple upside-down cake with plates, forks, and napkins beside it. Julia sat at the table, apparently guarding the cake.

"Thank goodness you're home," Julia said. "If you hadn't come soon, Tony would have eaten the whole thing."

Antonio led the gathering to the table, where they shared the still warm dessert while the young people gave Tony and Julia a play-by-play description of the game.

"Sounds like we missed a good one," Tony said.

"If we hadn't missed the game, you would have missed the cake," Julia said.

"Good thing we missed the game," he replied, smiling at his wife.

After they'd finished eating, they watched a 1940s western together, laughing at the stilted dialogue and outdated costumes.

"These cowboy movies sure would turn out differently if Indians wrote the scripts," Tony commented.

Juan agreed. "Just once, I'd like to see an Indian be the sheriff."

"Oh, Juan," Neyse moaned. "Won't you ever give up

that political bit?"

"No," he said, his voice no longer kidding.

"Well," Tony said, "I think I've had enough of this movie. Goodnight, kids."

"I'll join you," Julia said. "Goodnight," she told the others.

Paolo put his arm around Erin's shoulders and rested his head on the back of the couch. Every now and then Erin noticed that he was singing under his breath, but it took her a while to realize what it was.

"Is that from the dance?"

He nodded, looking almost embarrassed. "It takes a while to get it out of my system."

She wanted to ask what the words meant, but she knew that she mustn't. Even if they took the same road sometimes, there would always be that distance.

When he was leaving, Erin went outside with him. He kissed her, leaning against the wall on the front porch and pulling her close to him. She felt his excitement and the fear came back to her.

"You're very special to me," he whispered.

"I'm glad."

Glad, but still a little scared.

CHAPTER 13

"Now, tell me again why it's called Gallo?"

Neyse made a face. "Okay, but take notes this time."

Erin stuck out her tongue at her cousin.

"It's a feast day."

"Come on, Neys, I know *what* it is."

"It's a feast to celebrate certain saints—you know, Catholic saints."

Erin smiled patiently. "Yes, I know. Why is it called Gallo?"

"*Gallo* means cock in Spanish," Neyse continued as if she were speaking to a five-year-old. "A fighting cock."

Erin huffed in exasperation, walked out of the kitchen where the two girls were making their lunches, turned on the television in the living room, and sat down in front of the set.

Neyse began laughing.

"*Hin'a, hin'a,* I'll tell you."

Erin pretended not to hear her, so Neyse went into the living room and stood directly in front of the TV.

"They used to have cock fights on Saint's Day, a long time ago. Cock...Gallo...get it?"

Erin continued to stare toward the TV.

Neyse put her hands up under her arms and flapped her "wings."

"Bawk, bawk, bawk," she cackled.

Erin tried, but she couldn't contain her laughter.

"Chicken seems like a pretty stupid name for Saint's Day."

Neyse shrugged her shoulders.

"I take it they don't have cock fights any more."

"Right."

"What do you do for Gallo these days?"

"Families who have members named after the saints, like John, James, Paul, Lawrence or Ann, make feast. The village goes to those homes to celebrate."

"What else?"

"The families throw things, like food and things."

"They what?"

"They throw. That's what it's called."

"Throw them where?"

"Off the roof."

"*What?*"

"The family gets up on their roof and throws things down to the people."

"Why?"

"Good grief, Erin, I don't know why. Why does Santa come down the chimney? It's a custom. It's lots of fun, a real free-for-all. Everybody loves it."

"Okay, if you say so."

They ate their sandwiches, ignoring the program on the television.

"Off the roof," Erin said, sounding perplexed.

"Down the chimney," Neyse replied.

Then they both laughed.

"Will you have Gallo here?"

"No, but Aunt Victoria will. It's Paolo's Saint's Day."

"Will there be a dance?"

"No, but there will be foot races in the plaza. Some

pueblos even have a carnival, but we don't."

Erin was pleased to hear that news, although she refrained from saying so.

"It does sound like fun," she agreed at last.

They hadn't done much since the day of the baseball game, more than a week before. Erin was delighted to know there would soon be another celebration.

She was glad, too, that Paolo would be involved. She'd seen him only twice since the game: once at the lake and once at the rec center. He'd gone on a trip with Ramon the past weekend and she hadn't heard from him since his return.

She adored her cousin, but what she really wanted was to see more of Paolo.

When they'd finished lunch, Neyse said, "Let's go to the post office and see if we have any mail."

"Okay. Do you think your mom has anything she'd want us to drop off?"

"If she had, she would have taken it into town."

"When do you think she and your dad will get home?"

"Not until early this evening, I'm sure."

They walked down the road to the small adobe building that was the pueblo post office. Puffs of dust squirted out from under their feet as they walked, the ground hot and dry. Erin wore a pair of cut-offs that no longer fit Carlos, glad to give her favorite white shorts a day to be washed. Neyse also had on cut-offs, along with a red halter top, very much like the blue one she'd lent Erin to wear.

Neyse greeted the woman behind the counter, told her their box number and received in return, a sizable stack of mail, including a Penney's catalog.

She sorted through it.

"Here's one for you," she said, handing Erin a pale lavender envelope.

"It's from my mom," Erin said as she tore open the flap. She read it aloud as they walked back home.

"I suppose I should write her," she said, refolding it. "Since she's on a computer all day long, the last thing she wants to do at night is to email. She says it isn't a real letter unless you write it with a pen. And don't even get her started on texting."

"And then there's you," Neyse teased.

"Not so much these days," Erin replied thinking of her conversation with Paolo.

They spent the afternoon at the dining room table, Erin composing a letter home, Neyse looking through the colorful Penney's catalog, interrupting Erin frequently to show her a cute sweater or a funny-looking dress. They snickered at page after page of items like the "fanny lift brief" and "cinch-em-ups", as Erin's father called bras.

"You know," Neyse said, putting the book away, "the summer is more than half over."

"I know," Erin said. "I'm trying not to think about it."

"I don't want you to go home."

"Me either."

"I wish you could spend a whole year here."

"That would make for a long commute to school."

"Well then, will you come back next summer?"

"No. Next year it's your turn to come to California."

"I guess that's only fair."

Just then the front door flew open and Carlos came bounding in. "Hi!"

"Hi to you," his sister said. "What's up?"

"On my way to the rec center," he called over his shoulder. "Luis and I are going," he yelled from his room.

Moments later he came out, dressed in swim trunks and carrying a towel. "Tell Mom I'll be home later."

He slammed the door on the way out.

"He's like a Pony Express rider," Neyse said. "Stops just long enough to change horses."

"It seems strange that your mom and dad aren't home yet, doesn't it?"

"It's only five," Neyse said. "They don't go into town alone very often and when they do, they usually make a day of it. You know, go out to lunch and all."

When Tony and Julia finally returned, the girls learned that they hadn't been alone after all.

"We ran into Victoria at the new mall," Julia told them, "and went along with her to buy for Gallo. We found some wonderful things, not too expensive and yet, unusual."

She chattered on as she fixed dinner and Erin thought, *how young she looks*. It was almost as if she were a teenager. All during dinner, Julia laughed and teased Tony. Erin had not seen her aunt like that in all the time she'd been there. Tony noticed the change, too.

"A day alone with your wife makes for a happy woman," he said, giving Julia a squeeze.

She smiled. "It doesn't take much."

They spent a quiet evening. The girls' writing bug infected Julia. Erin wrote a letter to Scott and started another to Taylor. "She'll be shocked," Erin laughed. "I'm not even sure she can read actual handwriting." Neyse wrote Erin's family, their grandparents, and a pen pal. She made delicate drawings along the borders of her letters.

As she sealed and stamped envelopes Neyse said, "We should be getting a bag full of mail in the next month."

"Not likely," Julia said with a wry smile. "Hardly anyone, me included, takes the time to write letters much any more. Your mother is the star exception to the rule," she said to Erin. "It's a real shame, because writing has become a lost art. People either phone or text or use email. Nobody writes with a pen on stationery. In fact, I've heard that some schools have stopped teaching cursive writing. Imagine! How will the upcoming generations read the important historical documents if they don't know cursive?"

"I love to get mail," Erin said enthusiastically.

"I do, too," Julia agreed. "I just don't take the time to

write and send letters to begin with."

I'll never stop writing letters, Erin promised herself silently.

The next few days sped by.

The morning of Gallo, the girls slept in later than they'd intended. Erin was surprised at the time when they finally got up. She still wasn't entirely comfortable with the relaxed pace of the pueblo, often feeling that she should be "busy doing something."

They fixed themselves toast and juice and sat at the table trying to wake up.

Julia and Tony came in the front door, each with an arm full of vegetables.

"It smells wonderful outside," Julia said. "Everybody's baking bread."

"Were you down at the field?" Erin asked.

"Since early this morning," Tony answered. "You girls haven't been keeping up with your weeding."

Neyse hung her head. "Sorry, Dad."

"We went by Victoria's," her father said, changing the subject. "I had to try out her prune pie and see if it is as good as your mother's."

Julia poked him in the ribs. "Considering the fact that Victoria taught me to make prune pie, hers shouldn't taste the least bit different from mine."

"I'll admit that they're subtle differences."

"Like the shape of the pan she uses," Neyse teased.

"Where is Carlos?" Erin asked.

"Where else?" Tony answered. "At Victoria's."

"He might as well move in," Neyse commented. "He spends all his time there."

"Since you have Erin to be with, he's probably spending more time than usual with Luis," Julia said. "It works out fairly evenly when you consider how much time Paolo has been spending around here lately."

Erin tensed, but when she looked at her aunt, she saw a

broad smile and knew that she did not disapprove.

"Are you girls going to get dressed?" Julia asked.

Neyse shrugged. "Eventually."

Erin nudged her cousin. "Come on, lazy bones. Let's get going or we'll miss the races."

"Oh, all right."

They showered. Erin washed her hair, blew it dry, and then curled it with a curling iron.

Neyse watched her and smiled. "You certainly are going all out."

"It's his Saint Day. I want to look nice."

She put on a delicate, white cotton sundress, trimmed with white lace and ribbon, and tied a white ribbon in her hair. The stark white made her skin look especially dark.

Neyse chose a gray-blue print skirt and blouse, also trimmed with lace and ribbon. "I don't usually get this dressed up for Gallo," she said.

"Oh. Well, should we change?" Erin asked, looking disappointed.

"No. We can wear anything we want."

Erin was glad. She wanted Paolo to see her dressed up since he never had before.

"Would you French braid my hair?" Neyse asked Erin.

"Sure."

Neyse sat on the floor, Erin on the bed, while the latter braided the heavy, dark hair.

"Your hair is so thick compared to mine," Erin said unhappily.

"But yours curls and mine won't."

"You can't expect to have everything," Erin told her.

"I wasn't the one complaining."

Tony knocked on the door. "You two ready yet? It's past eleven."

They each took one final look in the mirror and then nodded at each other.

"We're ready," Neyse said.

The plaza was crowded with villagers, some participating in, but most watching the foot races. Tony found a spot, for the moment out of the sun, where they could see fairly well.

"Can anyone race?" Erin asked.

"Anyone who wants to. They're grouped according to age."

"Have you ever run?"

"Sure. And I've won. Once, I won a watermelon. But I dropped it on the way home. I was only six. Ants were all over it in seconds."

Erin shuddered, thinking of the crawling insects that were everywhere on the pueblo.

The elders called for another group, and a horde of teenage boys gathered at the starting line. A large sack was placed at the opposite end of the plaza.

"What's in the bag?" Erin asked.

"Flour, probably. Or maybe beans. Mostly they race for food. It's practical. Everybody needs food."

Tony overheard them and added, "Besides...one size fits all."

Erin laughed with her uncle, reminded of how very handsome he was when he smiled.

They watched three races and although Erin recognized several people, she didn't see the one she was most interested in seeing.

"What happens when the races are over?"

"We go to the houses of the people who are going to throw. The elders know who they are and lead everyone from house to house."

"Oh."

"Would you rather go to Victoria's right away?"

Erin hesitated.

"That was a stupid question. Come on, let's go."

"Is it okay?" Erin whispered.

"Of course. Why not?"

"See you at Aunt Victoria's," Neyse told her parents.

Tiya greeted them with enthusiasm and Erin patted his head. Paolo was busy helping Ramon carry bags of things up to the roof as they walked into the yard. He smiled at Erin and Neyse.

"Hey, how about bringing me a Coke?" he called to them.

"Sure," Neyse replied. "Erin would be glad to."

"Neyse," Erin said to her with not quite believable impatience.

Erin followed Neyse into the house and into the kitchen where Victoria was busy at the stove.

"Hi, Auntie. Need any help?"

Victoria kissed her first, and then Erin. "I'd love it. The boys are all at the races except Paolo and he's not much help in the kitchen."

Felicia ran in from her room and hugged the girls. "Are you going to stay?"

"Yes," Neyse assured her.

Erin got Paolo the Coke he had requested and took it out to him.

"Do you want anything, Uncle Ramon?"

"Not right now, thanks. Did you go to the races?"

"For a while."

"Too hot to stay?"

"No," she said, flustered by his question and not knowing quite what to answer. "Neyse thought Aunt Victoria might like some help."

Erin was amazed at the number of things that they were taking up to the roof. *It must have cost a fortune.* She remembered hearing Julia mention how generous Indians were. *No wonder they live so simply if they give this much away.*

Reluctantly, Erin went back into the house, where Neyse was cutting slices of still-warm bread.

"What can I do?"

"How about finishing the salad while I go and comb my hair?" Victoria asked.

"Sure."

Victoria handed her a knife.

"How soon will the people get here?" Erin asked her.

"Pretty soon. I think we're about third on the list."

The girls finished the salad and cleaned up the few remaining dishes. When Paolo and Ramon came in at last, Neyse poured them each a glass of iced tea.

"The throwing has started. It shouldn't be long now," Paolo said. He gave Erin an admiring look and a brilliant smile. "This tea is perfect," he said, clearly meaning that she, and not the beverage, pleased him that much.

Tony, Julia, Carlos, Luis, and Alonzo arrived shortly.

"Can Carlos help us throw?" Luis asked his father.

"Of course."

The boys all ran outside and scrambled up on the roof where they sounded like a herd of runaway reindeer.

Victoria and Felicia emerged from the bathroom, the young girl's hair French braided exactly like Neyse's.

"My, aren't you beautiful," Ramon told her and she giggled with delight.

When they heard the voices of the crowd gathering outside, Ramon stood. "Here we go," he said.

He and Paolo went up on the roof with the younger boys, as villagers surrounded the house. Erin went out to watch. Cups, towels, soap, can openers, cookies and crackers showered down from above, as people yelled and laughed and scrambled to catch them. Slowly Erin was drawn into the crowd, not knowing what she would do with any of the domestic treasures, but eager to catch something anyway.

One woman had a scarf around her neck, an apron around her waist and three bags of cookies under her arm. The small children scampered about, trying to anticipate where the next item would come from. They appeared as pleased to catch a box of laundry soap as they did to catch a box of cookies. Occasionally a name was called when a gift seemed particularly suited for a certain person.

Erin watched the increasing excitement.

Suddenly she heard Paolo yell, "Erin!"

She looked up just as Paolo threw the entire contents of a bucket of water down from the roof, hitting her full in the face. The crowd howled. Erin looked up at him as he stood grinning down at her.

She was drenched, head to toe, completely soaked, her dress sticking to her legs, her hair slicked down against her head like a peeled onion.

Paolo laughed.

She didn't.

She backed away from the house, and away from the crowd. She turned and walked out of the yard, down the road toward home. She heard the celebration continue at the same pitch as if she'd never been there. When she was out of sight of Victoria's, she began to run, tears streaming down her face.

She went around to the back when she got home, kicked her muddy shoes off at the door, and walked quickly into the bedroom. She found the shorts and tee shirt she'd had on earlier and went into the bathroom. She pulled off her clothes and turned on the water as hot as she could stand.

She stood in the shower, under the pounding water and cried. *Why? What had made him do that*? He'd made a fool of her. *At least Neyse didn't see it*, she thought with some relief.

She rinsed her hair and scrubbed her face, trying to wash away the indignation. Erin wasn't sure whether she was more angry or hurt, but she knew she didn't like the feeling, whatever it was. It had changed little by the time she emerged, dressed in shorts, a shirt, and a blue towel wrapped around her head.

She went into the kitchen to get a drink and was startled

to see Paolo sitting at the dining room table. He stood up and walked toward her.

"Erin."

She glared at him.

"I'm sorry, Erin. I didn't think you'd be upset."

"You didn't?" she snapped.

"No. It's part of Gallo. Everybody throws water."

"Everybody?"

"Every family who participates."

"Couldn't you have thrown it on someone else?"

He put his arms around her and tried to pull her to him. "Do you want me to treat you like an outsider?"

She pushed him away roughly. "If being an outsider means I don't get drowned and I don't get whipped, then I'll gladly be an outsider!"

"Don't get whipped? What are you talking about?"

"The River Men, that's what I'm talking about."

"And what do you know about the River Men?" he demanded.

"I know that they whip children, which is cruel and inhumane."

"Listen," he said in a voice which sounded almost threatening, "if you want to talk cruelty, let's talk about what *your* people have done to Indians in the past two hundred years: forced marches in freezing weather, denial of citizenship, destruction of homes, hunting grounds—in some cases, annihilation. Do you really want to talk about inhumane?"

"*I* didn't do that to the Indians," she screamed.

"And *I* didn't whip any children," he yelled back.

She began to cry.

He came to her and put his arms around her. "We're different. I told you that. You wouldn't listen."

Too different, she thought.

He leaned down and kissed her.

"Throwing water is part of Gallo. I thought Neyse

explained it all to you."

"I would rather have gotten a can opener."

He chuckled. "You're beautiful," he murmured softly and kissed her again. He held her tightly. She felt the anger melt away.

"I'd like to stay here with you, alone. But I think that we'd better go back."

She put her hand to the towel on her head. "Like this?"

"I think you look beautiful."

She smirked.

"Just leave it. It'll dry."

She thought about it and then she shrugged her shoulders. "*Hin'a.*"

She pulled the towel off, walked into the bedroom, and combed her hair.

Paolo came in behind her and put his arms around her waist. He nuzzled her and whispered, "I really would like to stay here."

"So would I," she agreed. "But I don't think it's a very good idea."

He kissed her on the neck.

"Let's go, okay?" she said.

"Okay."

Tiya jumped up the moment Paolo opened the door, and ran ahead of them down the road. Paolo took her hand and held it until they reached his house.

People were everywhere, eating, talking, laughing and examining their catch.

Neyse's mouth dropped open when she saw Erin.

"What happened? Never mind, I can guess. Paolo, you didn't."

He grinned, sheepishly.

"You're *kotra auksi.*" She looked at Erin and explained, "That means he's mean."

"It wouldn't have been such a shock to her if you'd explained Gallo properly."

"Me? Don't blame me!"

"*Tuu'ma,*" Paolo said. He looked at Erin. "That means I'm kidding," he told her. It was the first word he'd translated for her and it felt like an extraordinary gift.

Erin interrupted the argument. "You two can fight it out if you want to, but I'm hungry and I'm going to eat."

Paolo and Neyse watched in amusement as she found herself a place at the table, sat down and served herself a steaming bowl of red chili stew.

"I think she's over the worst of it," Neyse laughed.

CHAPTER 14

Neyse sat on the couch staring gloomily out of the window. The sky was dark and a heavy rain pelted the ground.

Erin shook her cousin's leg gently. "Come on, Neys, don't be a grump. We can go into town anytime. Don't let the rain spoil the whole day."

"Sometimes I hate this place."

"What?"

"You can't just go somewhere. You can't just go downtown or to a movie or shopping. You have to drive for an hour to do *anything*. It's such a big deal. California is so much better. We did whatever we wanted, whenever we wanted."

"Right. It never rains there, either. And money grows on trees. We can go into town next week. No problem. Let's do something else today, instead."

"Like what, for instance?"

Erin thought for a few minutes. "We could read."

"Read, shmead. We read yesterday. I want to *do* something."

"We could bake a cake or a pie."

"That sounds too much like work. I want to do something fun."

"Neys, you're not making this very easy."

Julia came into the room and put a load of books and papers on the dining room table. "What have you two decided?"

"Nothing," Neyse said.

"But we're working on it," Erin added brightly.

"Well, I'm going to work on lesson plans. That's always a good project for a rainy day."

Neyse harrumphed.

"I hope the storm isn't going north," Julia said, mostly to herself.

Tony and Carlos had gone to the Navajo reservation to the northwest to do some trading.

"Don't worry about Dad," Neyse reassured her. "He'll be fine. I'll bet he uses the storm as an excuse to stay over."

"That'll be fine as long as he calls me and tells me."

Erin perked up with an idea. "Why don't we go to the rec center? In fact, I challenge you to a game of pool. Whoever loses has to buy lunch for the winner when we go to town."

Neyse considered the challenge and then with a confident grin said, "You're on."

"Great," Erin said with pleasure.

"What do you know about pool, anyway?" Neyse asked.

"Not much, other than what I've learned at my grandparents'. They've had a table in the family room since I was about four."

Neyse groaned. "You're a hustler!"

"Not really. Besides, you agreed quickly enough. You must be good at it."

"Beats everyone in the family," Julia said.

Now Erin groaned.

"Will you take us, Mom?"

"I suppose, provided you stay long enough to make it worthwhile."

The girls agreed. "You won't have to come get us until dinner time," Neyse assured her. "And then, only if we don't get a better offer."

"Think that Juan might come by?" her mother asked.

Neysc smiled. "With any luck."

The two hurried into their bedroom.

"What should we wear?" Erin asked as they went through their clothes.

"Jeans, what else?"

"I sure am tired of these same clothes," Erin complained.

"If you were a reasonable size you could wear something of mine."

"What's the matter with my size?"

"Nothing. I'm just jealous." Neyse replied.

"Paolo says I'm too skinny."

"He actually said that?"

Erin nodded.

"What a dope. You're perfect. You look terrific. Lots of the young people here are overweight. Part of it is just our build, but Mom says it's mainly because the diet here is so bad. Indians eat too many sweets."

"You do have dessert a lot more often than we do at home."

"Mom is strict compared to some. Lots of kids eat junk food all day long."

Erin wondered if Paolo would blame that on the white man, too.

She put aside her thoughts and changed into jeans, a red and white striped shirt, and tennis shoes. Neyse also changed into jeans and snatched a blue tee shirt of Erin's out of her drawer.

"I'm not tired of your clothes," she laughed.

Julia ferried them to the rec center.

"We'll call you if we need a ride home," Neyse told her before they dashed for the entrance.

A surprising number of young people were there, which meant Erin and Neyse had to wait to get a table. They watched the other players and analyzed their styles quietly.

When their turn finally came they decided to play Eight Ball. Although they were fairly well matched, Neyse had a slight edge and won the game.

"Two out of three?" Erin said quickly.

Neyse nodded.

They played another game that Erin won. Several boys had gathered to watch them, making appreciative comments about their ability.

"Two to one on the blonde," a familiar voice said.

They turned to see that Juan had joined the gallery.

"Thanks a lot," Neyse said, making a face at him.

"She seems to have a natural talent," he observed with a grin.

"We'll see," Neyse told him.

The final game was over quickly, Neyse taking an early advantage and finishing with only two errors.

"Nice job," Erin complimented her. "I'll save up for your lunch."

"How are you at Ping-Pong?" Juan asked Erin.

Erin grimaced. "Lousy."

"What about a game, Neys?"

"Okay."

Erin sat and watched them hit the tiny white ball back and forth across the table.

"Good shot," she complimented Juan as he sank two balls.

"Nice going," she told Neyse when she did the same.

When they finally quit Erin exclaimed, "You're both good but Juan is terrific."

He smiled. "I know."

Neyse elbowed him in the ribs. "It's all he ever does up

here. Very limited skills."

"What about some lunch?" Juan suggested.

"You treating?" Neyse asked him.

"In California, the winner always treats," Erin commented.

Juan pulled a handful of change out of his pocket. "twenty-five, thirty-five, forty..."

Neyse laughed. "All right. We'll get our own."

They bought hot dogs, Cokes, and chips at the tiny snack bar and found an empty table near the windows. The rain still poured down outside.

"I'm surprised Paolo isn't here," Juan said, ripping open a bag of corn chips.

"I thought he was going to work for Mr. Montoya today," Neyse said.

"I doubt that they'd work very long in this rain," Juan observed. He looked at Erin. "You're not saying much."

She blushed.

"It's okay," he told her with a smile. "I like him, too."

Erin was curious to know if Paolo had said anything to Juan about her but she was unprepared to ask. He didn't offer any information so she said nothing.

When they'd finished their lunch, Juan challenged Neyse to another game of Ping-Pong.

"Why don't we do something Erin can do with us?"

"Don't be silly," Erin said. "I want to check out the library, anyway."

"Are you sure?" Neyse asked her.

"Yes, I'm sure. I don't need to be entertained."

While Neyse and Juan went to wait in line for one of the Ping-Pong tables, Erin wandered by the small, glassed-in office and read the announcement for fall classes at the center. Offerings included calligraphy, stained glass, jazz dance, pottery, creative writing, gymnastics, and leather working. It looked like an interesting variety to her, but she couldn't help wondering how many villagers actually participated in the classes.

She walked along next to the shelves of the library, a small area in the far corner of the center, and read the book titles. She recognized some of them but many of them she did not. There were selections from a wide range of subjects: literature, poetry, art, history, biography, gardening, business, psychology, and even a few cookbooks. One small space was devoted to younger children's books. The shelves were low and the floor carpeted, inviting even the smallest reader.

She browsed through several books, making mental notes of titles of those that interested her. *If only I were going to be here a little longer.* She never had time to read all the books that interested her. She finally selected a volume of Robert Frost poems, taking it to a comfortable chair near the window and within only a few minutes, she was completely absorbed in it.

She felt uneasy, all of a sudden. She had a sense of someone watching her. Looking up, she saw Paolo, his eyes riveted on her.

"How long have you been there?" she asked, feeling her cheeks turn red.

"Not long enough."

He stood at the end of the bookshelves, his arm resting on the books at the top, his chin on his arm.

She closed the book, a finger between the pages to mark her place.

"You don't have to stop."

"I'm not going to sit here and read while you stare at me."

He shrugged. He walked over, pulled a chair up next to her and sat down, careful, Erin noticed, not to touch her.

"Did you work for Mr. Montoya?"

"Yes. Until the rain was so heavy we could hardly see in front of us."

"Does it usually rain this hard in the summer?"

"We always have summer storms, but they don't usually last this long. They're generally over in a couple of hours."

He studied her for a moment. "Do you read a lot?"

She nodded. "I love to read. Books, that is. I just can't get excited about a Kindle. I like the feel of a book in my hands. And I love old books. My mom collects old children's books. The illustrations are charming...gentle and romantic...very different from the psychedelic colors and hyper style of so many books today." She tilted her head. "What about you?"

"I enjoy it more than I used to. It felt like too much work when I was younger. Now I enjoy reading, especially history."

"U.S. or world?"

"Both."

We're having a real conversation. She smiled at him. "This is nice."

He smiled back and nodded. "Read to me," he said.

"Seriously?"

"*Ha'a.*"

She paged through the book and chose the poem entitled "Acceptance." He watched her read, the words serenaded by the falling rain. When she finished he said, "Do another."

She flipped a few more pages and then selected "Acquainted with the Night."

When she'd finished he asked her, "What made you choose those two?"

"The first because of the title and the second because of the rain. I love rain: I love the sound of it, and the smell of it, and the feel of it. I even like walking in it."

He smiled. "I'm not surprised. I do, too."

Just then Juan and Neyse walked up. "I see you found her," Juan said. "Did you ask?"

"Ask what?" Erin questioned.

"We thought we'd play doubles in Ping-Pong," Juan said. "What do you say?"

"I don't think so," she said, shaking her head.

"Oh, come on, Erin," Neyse said with a teasing grin. "Don't let a little rain spoil the whole day."

"But I'm so bad at Ping-Pong."

Juan put his hand on her shoulder and winked at her. "With me for a partner, you can't help but be good."

She had serious doubts about that but as it turned out, she did quite well. At least they didn't lose every game. Juan encouraged her with well-chosen words while Neyse and Paolo hit easy shots to her.

After chasing the ball for what felt like the twenty-fifth time, she returned to the table with a grin. "I'm very good at jacks," she said.

Paolo frowned. "What is 'jacks'?"

Neyse couldn't repress a giggle.

Erin shook her head. "You never quit teasing, do you?"

He smiled at her. "*Dzah.*"

Juan looked at the clock above the office door. "It's after six. Do you two want a ride home?"

"Sure," Neyse agreed. "Mom'll be happy not to have to come for us."

"My dad took his car this morning so I've got the pick-up," Juan told them.

"I'm not going to ride in the back," Neyse said firmly.

"You don't *have* to," Juan said, sounding like he'd said it many times before.

Neyse's comment reminded Erin of hearing about the way Indian women used to ride in the back of trucks on the pueblos. Neyse had explained that it was the custom for them to do so, even in winter, because the men always drove.

Erin considered how white men had traditionally treated women like dolls. They never would have allowed a woman to ride in the back of a truck, exposed to freezing temperatures, while they were warm in the cab. Yet feminists complained about that treatment, calling it chauvinistic. Erin wondered if men didn't get confused sometimes, trying to figure out *what* women wanted.

There was certainly no doubt what Neyse wanted.

The four of them ran out to the parking lot, shielding themselves as best they could with their arms. They squeezed into the cab seat. Juan started the motor, and eased the truck out of the parking space.

"Why don't you guys stay for dinner?" Neyse asked.

Juan hesitated.

"Oh, come on. We can fix it for Mom. She's been doing lesson plans all day. I'll bet she'd love to be waited on."

"What are you having?" Paolo asked.

"Whale," Erin answered quickly.

"Are you sure you'll have enough?" he teased.

Juan was disappointed when Julia told him that she was fresh out of whale and that he'd have to be content with cheeseburgers. The young people fixed the meal, insisting that Julia relax in the living room on the couch.

As they sat down to the table, the phone rang.

"Bet it's Daddy," Neyse said.

Julia answered it and nodded to Neyse to indicate that it was. After a brief conversation she hung up. "He's staying over at his cousin's, just as you predicted."

"What a shame," Paolo said.

They all looked at him questioningly.

"I hate to see him miss out on these whale burgers."

After dinner, they sat in front of the fireplace, enjoying a fire the guys had built while the girls did the dishes. Julia said she wanted to finish the book she was reading and had gone into her bedroom.

"It seems strange to have a fire in the middle of summer," Neyse said.

"Today didn't seem much like summer," Erin replied.

"Summer is almost over," Paolo said heavily.

Erin felt colder just hearing him say it.

"We haven't gone up to the old village yet," Juan complained.

"We've been waiting for Neyse's ankle to get stronger," Paolo said. "Think you could make it?" he asked her.

She shook her head no. "It doesn't hurt anymore, but I still get tired of walking pretty quickly. I could never make it that far. Maybe you guys should go without me."

Erin and Paolo looked at each other. Juan immediately said, "Why don't you two go?"

"I'd show her the ruins if she'd like to see them," Paolo said. He looked at Erin, uncertainly.

"There isn't a tram, is there?" she asked hopefully.

Neyse giggled.

"Hardly," Paolo said.

"I have to admit that hiking isn't my favorite thing to do but if it's the only way to get there, it's better than not seeing it at all."

"Since the rain will cool things down, the sooner we go the better. We don't want to hike it in the heat."

"Why not go tomorrow?" Neyse encouraged them.

"Might be pretty wet," Juan cautioned. "Day after would be better."'

Paolo looked at Erin. "Is that all right with you?"

"I guess so."

He nodded. "Day after tomorrow it is."

CHAPTER 15

The next day was cool and cloudy, threatening into the late afternoon, but never actually turning to rain. Paolo called Erin after dinner.

"Are we still set for tomorrow?" he asked.

"Have they built the tram yet?"

"No, but we've been commissioned to decide where to put it."

"Right next to the restaurant at the top."

When he laughed, she felt a familiar delight; her joy was now his joy.

"Speaking of food," he hinted.

"I know, you want me to fix a lunch."

"If you'll fix it, I'll carry it."

"I suppose that's fair."

"I'll pick you up around eight o'clock, okay?"

"See you then."

The morning dawned, cool, clear, and slightly breezy, all but a few traces of the storm gone. Julia woke the girls, told them that Tony had already left for work and that she was headed to the market.

"Have fun today," she said, kissing Erin goodbye.

"Thanks, Auntie. I'm sure we will."

Erin dressed in jeans, a long-sleeved, pale blue pullover, tennis shoes, and socks. Neyse offered her a warm sweatshirt.

"You may not need it, but take it in case. Beats getting your beautiful white sweater dirty."

Neyse sat at the dining room table after they'd finished eating their breakfast, chattering while Erin packed the lunch. When she was done, she put it into a backpack.

"What have you got in there, anyway?" her cousin asked.

"Sandwiches, cheese, crackers, fruit, and cookies," Erin told her.

"You're only going for half a day, Erin."

"I know, but I may need to take a lot of nutrition breaks."

She was ready long before Paolo was due to arrive. She went to the bathroom three times in less than half an hour.

"What are you so nervous about?" Neyse asked her.

Erin shrugged her shoulders.

"You like him a lot, don't you?"

"I really do."

"Wouldn't it be neat if...?" Neyse began, but didn't finish.

Erin wondered if they'd been thinking the same thing. *Wouldn't it be neat if we fell in love, if I moved here, lived here?* But, as she thought it, she felt annoyed with herself. *You dummy: you're eighteen years old...well, almost. Why in the world are you thinking about marriage?* Her answer? *I don't necessarily want to, sometimes it just happens.*

"Erin. *Erin,*" Neyse said louder. "What are you day dreaming about?"

Her head jerked toward her cousin. "What?"

"You're a million miles away."

"Yeah, I guess I was." She changed the subject. "Are

you sure you don't mind us going without you?"

"Don't be dumb. Of course I don't mind. I've been to the ruins. Sure, it'd be fun for all of us to go, but it's silly for you not to go just because I can't. Besides, it'll be more romantic for you two to go alone."

Erin walked to the front window. "When is he going to get here?"

"Eight o'clock Indian time could mean anywhere between nine and noon," Neyse reminded her.

Erin shook her head. "That's one thing I could never get used to."

"Oh, you could...if you had to."

Never, she thought.

Neyse tuned in a morning talk show on the television but Erin couldn't even begin to concentrate on it. She was relieved when she finally heard the sound of a truck in the driveway. She grabbed the lunch and gave Neyse a quick hug.

"Tell your mom bye for me when she gets back."

"Erin, you make it sound like you're leaving for weeks. You'll be home for dinner."

"I know. I'll see you later."

She walked out the door just as Paolo was getting out of the truck.

"Ready?" he asked.

"For hours," Neyse said before Erin had time to answer. She'd come out on the porch right behind her cousin.

Paolo shook his head. "Shows poor upbringing," he said, looking worriedly at Neyse.

Erin surprised them both by throwing the backpack at him. He caught it easily and then added, "Guess she doesn't agree."

He opened the cab door for Erin and she climbed in. When he slammed it, the window shook in the door as if it might shatter. He popped down the lock.

"Bye, Neys," he said. "Be good."

"I'm always good."

Paolo walked around the back of the truck, put the backpack behind the driver's seat, and jumped in.

"Buckle up," he said, nodding at the seatbelt. When they were both secured, he backed quickly out of the driveway, creating a huge cloud of dust, as the ground was already dry after the rain.

Erin waved to Neyse who stood coughing in the yard.

Paolo drove along the road toward the dam, passing that turnoff as well as the road to the rec center. For some reason, Erin felt stiff and uncomfortable again, almost as if she didn't know him. She wondered if she'd made a mistake, agreeing to come with him alone. He looked comfortable enough, in a pair of faded cords and a plaid shirt with the sleeves rolled up. He seemed perfectly relaxed as he drove.

"You certainly are quiet," he said as they bounced along the dirt road a little later.

She looked at him and said the first thing that came into her head. "It's such a beautiful day."

He nodded. "It is. And it was clever of you to notice." Then he changed his voice to sound like a computer. "Now you can turn to page two of your dating manual."

She laughed.

"Are you nervous?" he asked in amazement. When she shrugged, he said, "You don't need to be. I'm the same wonderful guy who threw a bucket of water on you just last week. I haven't changed a bit."

They laughed together and the tension disappeared.

"Now, for the important things. What did you make for lunch?"

"Oh, the usual, sliced whale sandwiches."

"Is that all you know how to fix? Haven't you ever heard of peanut butter?"

"You mean piñon nut butter?"

"You're getting all too good at this."

"I was good at it before I ever got here."

"Everybody in California is a comedian, right?"

"Right. And it never rains there, either."

He frowned, not understanding.

"It's a long story," she said. "Say, where's Tiya?"

"Home. I wanted you all to myself today."

She wasn't sure whether to be pleased or nervous. At this point she could only wait and see.

They continued along the gradually narrowing road for several miles, drawing closer to the distant mountains. Erin knew exactly where they were when they passed the turnoff for Dixon's. Finally, Paolo slowed the truck and pulled off the road into a small clearing. He pointed to a mesa that seemed miles away.

"There it is."

"Doesn't look like much from here, does it?"

"We don't have to go if you don't want to," he said, sounding defensive.

"Don't get touchy. I want to. They may never get the tram built if we turn back now."

He pulled the backpack out from behind the seat and opened it. He took Erin's sweatshirt and put it, with his own, on top of the lunch. He carefully zipped it up.

"Ready?"

She hopped out of the truck. "Ready!"

The first half hour, the hike wasn't difficult. Paolo led her around bushes as best he could, and looked back at her frequently to make sure she wasn't winded. But as they neared the mesa, the terrain changed. The shrubbery was more dense and harder to navigate. Erin began to anticipate the climb awaiting her; she slowed down, forcing Paolo to do the same.

Instead of commenting on her slackened pace he said, "There's a very old story my father used to tell us, about a young woman who was kidnapped by the Navajos and taken far away to live in their village. One of the men who kidnapped her, took her for his woman and she lived with him for many years and even bore his children. But she was always unhappy. She hated her captors and longed to return to our people."

He stopped to catch his breath as they had started the long climb up the mesa.

"Over the years she was befriended by an old woman of the tribe. One day this old woman told her she knew how she could escape. 'Tomorrow when you go with your man to gather berries, he will ask you to wash his hair. Do so, and as you comb it dry, he will fall asleep. When he is asleep, you must slit his throat, take his scalp and wrap it in a cloth. Put the scalp inside your dress next to your heart and it will speak to you.'"

Erin groaned. He held out his hand to her and pulled her up the slope.

"Is this too gory for you?"

"Yes. But go on anyway."

"The old woman told her to take two horses and ride as fast as she could, switching from one to the other as the horses tired. The scalp, the old woman promised, would direct her toward our village. The next day, the captured woman did exactly as she had been told. Her hatred for the man who had kidnapped her, gave her the strength to kill him and scalp him.

"She rode hard and fast and had a considerable start before the other warriors found the slain man and started out after her. She rode all day and all night, the scalp telling her what to do."

Erin climbed, so involved in the story that she no longer noticed the steep hill.

"The following day, the Navajos were gaining on her as she neared our village. As she approached the fields, she saw an old man. She spoke to him but he was suspicious of her for he didn't recognize her. In those times, a stranger who spoke the tribal dialect was looked on with great distrust. She called him by name and he was afraid. He demanded to know who she was. 'Father, don't you remember me? I am your daughter who was taken long ago.' He was amazed. She told him quickly of her escape and of her pursuers. They hurried to the village to warn the others. When the Navajos arrived they were easily defeated by our warriors."

He stopped to catch his breath. "We're almost there."

She looked up. "Is it a true story?"

"So they say."

She shook her head. "How could she leave her children?"

"They weren't hers, they were his."

She thought about that but didn't understand how he made the distinction.

"Did they live in this village when it happened?"

"I don't know. I suppose they could have."

Paolo scrambled over the top edge and helped her up the final few steps. Erin had expected to find a village like Taos, perhaps in bad shape, but nonetheless, buildings. Her disappointment was great when, in fact, she saw nothing. Paolo walked on and she hurried to catch up with him. They walked for another five minutes before they came upon the ruins.

Erin saw long lines of rocks, stacked up two or three high, apparently cemented with adobe clay. They clearly had once been the walls of the pueblo, but now ruins were all that were left. Nevertheless, a strange, eerie quiet surrounded them and Erin felt as if she'd entered a holy place.

Paolo slipped off the backpack and set it near a wall that was about two feet high.

"Too tired to walk around?" he asked.

His voice split the silent air.

"No."

He took her by the hand and led her around the age-old remains. As they looked into the small, square rooms Erin tried to imagine people living there: cooking, eating, sleeping. She tried to see children playing, and growing up, high on the flat mountain.

"How did they keep the little ones from falling off the cliffs?"

"I suppose that they tied them to something."

"Like animals?"

"What do you think people who live on boats do with their kids?"

Erin shrugged. "The same, I suppose."

"Survival is not always easy or pleasant."

"They could have made little play pens for them."

"You mean little cages?"

"No!"

"I'm just saying it's a matter of perspective. Don't you see?"

She looked away from him, thinking. "I know you're right. It just sounded so barbaric."

"Without understanding, many things do."

Taking her hand, he led her back to where he'd left the pack, sat down next to it, and stretched his legs out straight in front of him.

"Sit down with me. We deserve a rest."

She stretched out beside him and leaned against the wall.

"Climbing wasn't as bad as I thought it would be. I loved hearing the story. Did your father tell you many stories when you were young?"

Paolo nodded. "Our culture has been passed from one generation to the next by telling such tales."

"Tell me another one."

"In a while."

He put his head back against the adobe and closed his eyes. She looked at him, fascinated by his dark, smooth skin and glossy black hair. She would have liked to reach out and touch him.

He opened his eyes suddenly. "What are you thinking?"

"Your hair is beautiful."

"Men don't have beautiful hair; Indian men, anyway."

"You do."

He leaned toward her and kissed her. She felt excitement in her stomach, an excitement she liked, but that scared her. She returned the kiss. He took their sweatshirts from the backpack and spread them open on the ground. He reached out, then, and pulled her down next to him, settling her head on his shoulder.

The sky overhead was a piercing blue. Puffy white clouds hung close enough to touch. They lay silently next to each other. She wanted it to last forever. She wanted to never leave.

Some time later, she woke. When she turned to look at him, he was smiling contentedly.

"I fell asleep."

"I know."

"For very long?"

"No, not long. I liked watching you sleep. You looked like a little girl."

"Did you sleep?"

"No."

He held her.

"Are you hungry?" she asked.

"Starving."

She sat up, rubbing her eyes. He reached into the backpack and handed her a sandwich. Erin took a bite of it but discovered that she had no appetite at all. Paolo opened a can of Coke and offered her a drink. It was still cold and tasted grand.

He kissed her again, drops of Coke on his lips. "You're not eating."

"I'm not very hungry."

"You'll need your strength for the trip back."

"I'm going to slide the whole way."

"It's been done before," he assured her.

Erin felt so utterly happy she thought she might float

away. She'd been happy before, so happy that she thought her heart might burst through her chest. But this was an inside and outside happiness. It was different.

He looked at her and smiled. "What are you thinking?"

She sighed deeply. "I'm so happy I could dance on a cloud."

His gaze bored deeply into her, thoughtful and intimate. He put down his Coke, reached out to her, and held her hands in his.

"Cloud Dancer."

She had to make herself breathe.

"It's a good name for you."

"Say it in your language."

He thought. "I can't, really. The closest would be *Henateetz Kamakas*."

He nodded, almost imperceptibly and repeated it. "Cloud Dancer."

"Do you have an Indian name?" she asked, then frowned. "Am I not supposed to ask that?"

"No, it's okay. My name is *Whooshka*."

"What does it mean?"

"It means Robin."

She repeated it softly.

"You are so beautiful…" He kissed her. "...and so special."

"Paolo, would you tell me something?"

"What?"

"Why did you dislike me so much at the beginning?"

He looked uncomfortable. "I didn't."

"When I first got here you certainly did. You were nasty to me from the very first day."

"I didn't dislike you." He was quiet for a while, unwilling to look at her. "I wanted to. I guess that's why I acted the way I did. I liked you the minute I saw you. I just couldn't admit it. After you went to Taos, I couldn't pretend any longer."

"You mean all the time you were being mean to me..."

"I was just trying to kid myself, or maybe convince myself."

"I couldn't understand you."

"I was confused, too," he confessed. "I had always hated outsiders. I was able to stay away from them and keep them away from me. Then, you came along. You didn't seem like the others. You seemed like—like me. I wanted to be with you and yet I didn't. I couldn't understand what was happening. It scared me. That made it worse."

She smiled at him, feeling very pleased. "You're learning that people are the same, no matter where they're from."

"No, Erin, we're not the same. We're different. It's important for you to understand that. You *will* understand it, one day. I like you, but we are very different."

He kissed her again, almost as if he were trying to heal the wound of truth.

She cuddled up against him, happily not hearing.

"Will you tell me another story now?"

He nodded. "Yes."

He thought for several moments before he spoke again.

"There was once a very poor family in the village. The parents were old and their only child, a daughter, decided that she would work hard to care for them. She gathered scraps of cotton that had been thrown away, combed them, spun them, and rolled the yarn into a ball. When she had enough yarn, she made a pair of footless stockings. When she'd finished them she gathered more cotton, worked it and made openwork stockings. She showed her parents what she had done and they were pleased.

"Then she decided to make a white manta. She gathered cotton, combed, spun, and rolled it into yarn. She used her loom to weave the big white manta. When she had finished weaving, she decided to embroider it with many colors. And so she did. Her parents were very happy.

"By now, she had grown to be a good looking girl and the young men came to her to ask her to marry. She told them, 'I take care of my parents and myself. I do not want to marry.'

"She continued working hard, making first a small white manta, and then a belt. Her father told her, 'When you finish the belt stretch it well, so they won't stretch you when you die.' This is the advice given to all Indian girls when they weave."

Erin didn't understand what that meant but she didn't want to interrupt the story, so she made a mental note to ask him later.

"Her parents sold the things she made and she went on, making more. The people of the village came to her and bought her garments. At last, everyone in the pueblo had a complete dancing costume of his own that she had made.

"The village people decided to have a great dance in front of her house, to see who she would dance with. They did so, but she sat weaving and would not even look up. 'Why do you all come to me? I am not the only girl in the village. I only want to care for my mother and father and stay where I am.'

"The young men decided to try to raise colored corn to attract her. Though they grew corn of blue, white, red, yellow, and dark red, she was not interested. 'I tell you I never want to marry. I make my own clothing and live well.'

"The boys decided they would not court her any more.

"Coyote heard all this and decided that he would trick her into going with him. First he found a branch of black currants. He went to the village to a house near where the girl lived. He began to dress to dance. With each piece of clothing he put on, he stamped his foot four times and said, 'Do I look pretty? Yes, I look pretty.'

"When he was all dressed he went to the plaza to dance. When the girl heard the singing she put down her work and went out. 'What a fine looking boy. I wonder who he is.'"

"But I thought you said it was a coyote," Erin interrupted.

"Yes, but Coyote is a master of trickery. When he was all dressed up, he looked to her like a boy."

"Oh."

"The girl saw the branch of black currants and asked him for them. Then she took Coyote to her house. The boys of the village were very angry for they had offered her much more and yet she had refused them.

"The girl and Coyote slept together that night, and soon she was ready to give birth. She had two children, half human and half coyote. Coyote and his wife and children went to his home far away. It looked like a mere hole in the ground but when the girl entered, it was a house as fine as her own. They lived there always."

When he finished, he looked at her, a sliver of a smile on his face.

Erin grimaced. "Half coyote?"

Paolo nodded.

"How awful."

"The girl had to be taught a lesson."

"And what was that?"

"Girls are expected to marry and have children."

"You aren't much for women's lib, then?"

"No," he said flatly. "Our women must reproduce or our pueblo will not survive. It is necessary. Whenever an individual's will is in conflict with the welfare of the pueblo, the individual must give in. It's been this way forever."

Erin bristled. "Sounds like an excuse to be old-fashioned."

"It is."

"Do you think it needs to be that way now?"

"Only if the pueblo intends to survive."

Since she could think of no way to argue with him, they sat quietly for a time.

"Are you ready to go?" he asked.

She frowned and got to her feet. “Not yet. I want to look around a little more.”

“The story upset you?”

Although it had, she shook her head. “It surprised me, that’s all.”

The sun was far down in the sky; the air had started to cool. Erin guessed it was around five o’clock, although she wore no watch.

They wandered around the mesa, picking up small pieces of rock and shards.

“I wonder what it would have been like to live here,” she mused.

“Unlike anything we know.”

“Could you live here...now?”

“I don’t really know. It wouldn’t be easy. I’m used to today’s ways: cars and telephones, televisions and running water. I don’t think I could give all that up.”

She was surprised that he was so candid.

“In a way, it would be like what Julia had to do,” she said.

He looked hard at her. “Yes, it would.”

“Where do you see yourself in ten years?”

He chuckled. “That’s a white man’s question; a question a person asks when he isn’t satisfied with his life; a person who wants more.” He looked hard at her. “I am satisfied with my life. I don’t need more of anything.”

She felt a little envious and a little disconnected. She felt the difference he’d been talking about.

They were both quiet as they looked over the edge of the mesa toward Dixon’s farm. The green orchard trees contrasted sharply with the desert surrounding them. Paolo put his arm around her and kissed the top of her head.

“We’d better go. I don’t want to be hiking after dark.”

“Okay,” she said, reluctantly.

He put the remains of the lunch and the two sweatshirts into the backpack and slipped it onto his shoulders.

"I wish we could just stay here," Erin said.

He took her in his arms. "So do I."

They kissed once more and then he led her back to the edge of the mesa. He started over the side saying, "There once was a deer with three fawns and a fox with three cubs..."

She followed, listening intently.

CHAPTER 16

"What did you think of the ruins?" Juan asked Erin that night.

He and Neyse had made popcorn and hot chocolate, which Erin and Paolo were helping to devour. Julia, Tony, and Carlos had gone to a movie in Santa Fe, so the four friends were alone for the evening.

Erin thought a minute, trying to find the right words to express her feelings. "It was so quiet that it seemed, sort of like a church, I guess. I felt as if I were walking on sacred ground."

"Sounds very serious. Did you have any fun?"

"Of course. Especially hearing the stories. Oh..." she said, smiling at Paolo, "and I have a new name."

He nodded.

"Cloud Dancer." She smiled with pleasure. "I wanted it translated but Paolo said it didn't work very well."

Juan considered it. "We could call you *Henateetz*. That means cloud."

"*Henateetz*. I think that's a good compromise," Erin agreed.

"I like it," Neyse said. "It fits you. Cloud Dancer."

"Who usually decides on your Indian name?" Erin asked.

"The grandparents or the godparents or one of the elders," Paolo told her.

"When?"

"When a child is baptized," he said.

"Usually it's done a few days after birth," Juan told her. "Corn meal is used in the naming."

"Corn meal?"

"Corn meal is used for many of our ceremonies."

"Go on."

"One person holds the baby and another holds a hand full of corn meal just under the baby's nose. When the baby exhales, his breath blows the corn meal into the air. Then he's taken outside, and given his name. The ceremony always takes place at dawn."

"Is that name on the birth certificate?"

"No. The parents give the child his legal name. This is more of a family name. In fact, some people have two or three Indian names."

"My mom has three," Neyse told her.

"Why?" Erin asked.

"Because three of our friends wanted to name her."

Erin wondered why she could be accepted enough to have three Indian names, yet not be allowed to dance.

"Seems to me it worked out perfectly," Juan said. "Paolo baptized her for Gallo and named her today."

"Frankly, I preferred the naming to the baptizing," Erin told them.

Paolo looked at her sheepishly.

"Speaking of baptisms," Neyse said, "it's getting close to our birthdays. We're going to have to plan a party."

"And go shopping," Erin added.

"One thing at a time," Neyse said. "Let's plan our party. Since my birthday is the eighteenth and Erin's is the nineteenth, what day is best?"

"What about the following weekend?" Juan suggested. "We can make a day of it."

Neyse got the calendar out of the kitchen. "Okay, Saturday, then?"

They all agreed.

"Fine. Now, what will we do?" she asked.

Erin spoke up right away. "As long as we make fresh peach ice cream, I don't care what else we do."

"I think we should go fishing," Juan said.

"I don't," Neyse replied.

Erin and Paolo laughed.

Neyse sat, looking at the calendar. "Erin, do you realize that you leave the following weekend? This summer has gone by so fast!"

"I know," she said, dolefully.

"I don't want you to go," her cousin said sadly.

"Better not get started on that," Juan cautioned, "or you'll be crying the rest of the night."

"Well then, let's plan the big day. Somebody say something," she begged.

"I've got an idea," Juan said. "Why don't you let Paolo and me plan your party? We'll surprise you both."

"That would be okay with me," Neyse said.

Erin hesitated. "I don't know. Paolo's idea of a party probably includes throwing things and water fights."

Paolo joined in after an encouraging nod from Juan. "I think it's a great idea. It'll be a birthday neither of you will ever forget."

"When you put it that way," Erin said, "it makes me even more nervous. I mean, I'll never forget Gallo, will I?"

Paolo moaned. "I wish you'd let *me* forget it."

"I want a real special birthday," Neyse said. "Promise it'll be fun?"

"It'll be fun," Juan assured her.

The two girls looked at each other, uncertainly at first, but finally they both smiled and nodded.

"Okay," they agreed.

"Good," Juan said. "Now, when are we going to town?"

Neyse flashed a look at Erin. "We're going in with Mom next week."

"Let's go together," Juan suggested.

"Oh, I don't know. We already made plans with Mom. What do you think, Erin?"

Erin merely looked at her with wide, uncertain eyes.

Neyse considered the idea. "I think it should be girls only."

"Talk about women's lib," Juan huffed.

"Oh, knock it off. We're going shopping. It isn't that we don't want to *be* with you, we just don't want to *shop* with you. You'd trail along behind us and be bored."

"Actually, I think it would be fun but if you don't want us to go along, we won't." He pretended, unsuccessfully, to sound hurt.

Erin looked at Paolo and caught him covering a yawn. "Are you tired?"

"After carrying you halfway up that mountain today? Of course I'm tired."

"Paolo!"

"It's okay, Erin," Neyse reassured her. "We know him too well to fall for that line."

"Believe it or not, I am tired," Paolo said. "I think I'll get going."

Erin was disappointed. Talking about the end of the summer made her realize how little time they had left to be together.

"Do you want me to walk you home?" Erin teased.

"Out to the front porch would be fine."

He took her hand to pull her up from the couch and kept hold of it as they went outside.

"Today was great," he said. "I'm glad we went."

"I am, too."

He leaned against the house and pulled her to him. "I'm

going to be pretty busy with Mr. Montoya the next few days. I probably won't have much time to see you."

"That's okay."

"Don't forget me."

"Don't worry," she said, and for the first time, she kissed him.

He caressed her, kissing her mouth quickly, then slowly and then quickly again. "We should have started this earlier," he whispered, his excitement obvious.

She tried to pull away a little but he held her tightly.

"*Henateetz*," he whispered, and moaned. "How can you make me feel strong and weak at the same time?"

She felt safe on the porch, but a little afraid that Tony and Julia might come home or that Juan would open the door.

"I really better go," he said at last. "I'll call you."

"Okay."

She watched as he walked across the yard, out the gate, and down the road into the darkness. Her heart was racing and she knew her face was flushed. She stood long enough to cool off before going back into the house.

Neyse and Juan sat on the couch, kissing.

"'Night," she said, and hurried into the bedroom.

By the time she took a shower, Juan had left and Neyse was ready for bed.

"Hope you didn't mind my saying the shopping trip was for girls only," Neyse said. "I adore Juan but I'm not ready to spend every minute with him."

"No, that's fine," Erin said as she climbed up to her bed. "Besides, men are awful shoppers. At least my dad and Scott are. They get tired fast and have no curiosity."

Neyse laughed. "Right. You know, it's funny how a man can climb Mt. Everest but he can't seem to keep up with a woman on a shopping trip."

They giggled together.

Neyse turned out the light, mumbled goodnight and fell

asleep almost immediately.

Erin lay awake for hours, the day wrapped around her like a warm shawl. It had been nearly perfect. She couldn't help wonder why Paolo hadn't pushed her when he'd had a perfect chance, alone on the top of a mountain. He'd kept his word. Was she disappointed that he had?

The weather had turned hot again by the time their shopping day arrived. Julia encouraged the girls to wear shorts having overheard them talking about dressing up.

"Besides," she argued, "if you're going to try on clothes, you'll be happier to have shorts on."

So, they wore shorts, all three of them.

"Do you have somewhere particular in mind?" Julia asked as they turned onto the highway toward Albuquerque.

"Old Town," Neyse said, "where else? It's beautiful, historic, and it has a ton of shops."

"And a million tourists," Julia complained.

"Oh, Mom, that's part of what makes it fun."

"All right."

Julia parked the car in a lot across from the plaza.

"It's just like Santa Fe and Taos," Erin said, "only this plaza is prettier."

"I told you it was my favorite," Neyse agreed.

They walked across the street and stood under the shade of the large green trees. The buildings were set around the square, the center of which was a small park with lush, green grass; large, leafy trees; a romantic white gazebo; and ornate wooden benches. Three of the streets housed shops; on the forth an old adobe church stretched the entire length of the block.

"What would you like to do first?" Julia asked. "Want to look around for a while and meet for lunch?"

"Aren't you going to go with us?" Erin asked.

"No. I have a few things I need to look for, alone. Why don't we meet back here around noon? That gives you almost two hours. You should be ready for a break by then."

The girls agreed. As Julia headed across the plaza, Erin took a sweeping look around the square. "Where do we start?"

"We can either go from store to store or I can take you to my favorite places first, and then see how much time we have left for the rest when we're done."

Erin took no time to consider the options. "Let's go to your favorite places first."

Neyse thought for a moment and then said, "I think my very favorite is the Christmas shop. Come on."

She led the way down the street by the side of the church to the alley behind. They entered the adobe building on the opposite corner to find Christmas in August. Ornaments of every shape, size and design, made of wood, glass, dough, metal, pottery, cotton, silver, and gold, were displayed. "Silent Night" was playing on the stereo system and despite the heat, Erin floated in a holiday mood.

Hanging from the ceiling were huge white snowflakes, crocheted from white cotton thread. Near a window, cut crystal glass ornaments caught the sun's rays and danced rainbows across the room. On one shelf that covered the length of the room were displayed every imaginable sort of nutcracker: a soldier, a drummer, a chimney sweep, a lumber jack, and a mountain man, each exquisitely detailed and colorfully painted.

"Look at the different crèche scenes," Erin cried as they went into the adjoining room. Elaborate nativity sets made of wood, marble, paper mache, and porcelain fascinated both girls. One was made of pottery in a typical pueblo design, simple and innocent in its execution.

"I want to get something here for my mom," Erin said. "She absolutely loves Christmas."

She looked carefully at each ornament and finally chose a delicate, gold-plated bell. It was paper thin and made a sound like angel wings when she rang it.

After she paid for the bell, Neyse led her to a shop nearby that specialized in kaleidoscopes. They were all works of art, each individually made by more than fifty artists.

"These are extraordinary," Erin exclaimed, looking at the many samples. "I've got to get something here, too." They were all so expensive, some in the thousands of dollars, that Erin could only afford a postcard. "Maybe someday I can come back for a real one," she said wistfully.

"Now," Neyse said, "for the candy shop."

"Mmmmmm," Erin said.

They looked in the windows of the stores as they walked the block and a half to the next of Neyse's favorite places. Erin ducked as they walked in the door, to avoid running into the many ribbons that hung from a piñata suspended from the ceiling.

The shop was magical, filled with candy, colorful boxes, baskets, stuffed animals, delicate pieces of china, and charming figurines. The piñatas were covered in pastel crepe paper and bouquets; pink, blue, green, and yellow ribbon hung down everywhere. Every nook was packed, every shelf covered. Even the walls and ceiling were used to display gifts.

"It looks like a fairyland," Erin said, turning round and round, trying to take it all in.

"And smells like heaven," Neyse added breathing in the blissful fragrance.

Erin looked in the glass display cases. "What are you going to get?"

"Orange cream with dark chocolate," Neyse answered. "That's what I always get."

"May I help you?" asked a crumpled looking, old woman with a sugary smile.

"I'd like two of those," Neyse said, indicating the dark chocolate, orange creams.

The lady put on a clear plastic glove and carefully

picked two of the hand-dipped chocolates, placing them in a small bag. Then she tied it with strands of curly ribbon in five different colors.

Erin decided on a dark chocolate butter and a vanilla cream. The woman put them into a bag, which she decorated in the same, precise fashion.

"Come again," she chirped after they paid.

"We will," they assured her.

"Where to now?" Erin asked.

"Anything special you'd like to see?"

"Some Indian things."

"Okay. Let's go look at the jewelry the artists bring here to sell themselves."

Just as in Santa Fe, artisans lined the sidewalk across the street from the Plaza, displaying silver and turquoise jewelry, belts, beads, and moccasins. Erin and Neyse walked from one blanket to the next, looking closely at the rings, bracelets, and necklaces. Erin lingered over a pair of delicately beaded moccasins.

"See anything you'd like?" Neyse asked.

"Yes," Erin said quietly, "but I can't afford any of it."

"Let's try the trading post," Neyse suggested. "Sometimes the things there are less expensive."

The garish displays of the "tourist traps" reminded Erin of stores on Fisherman's Wharf in San Francisco. Manufacturers wrote Old Town or Albuquerque or New Mexico on every conceivable object and offered it for sale to the tourists. Erin wrinkled her nose and said, "Ugh."

Once they were outside, Neyse looked around. "I know. Let's go to the basket shop. They've got really great stuff there."

They walked down the street, now crowded with people moving quickly in both directions. Neyse's shoe came untied and though she moved to the edge of the sidewalk to tie it, an overweight, overdressed woman nearly ran into her.

"Stupid Indians," she grumbled as she maneuvered her

tank-like body around Neyse and picked up her pace again.

Although her voice was low, Erin heard her clearly and looked, immediately for Neyse's response. The dark-haired girl appeared not to have heard the hateful remark. Erin started to ask her but decided not to. She was so angry that she wanted to run after the woman and smack her. *How could she talk that way, the fat cow?*

A hole had been torn in the day. A day which had started out perfect was now ripped. It hurt Erin. Even if Neyse hadn't heard it, Erin hurt for her.

The two girls walked the rest of the way to the basket shop in silence.

The large wooden-floored building made Erin feel as though she'd stepped back in time. The boards creaked with each step she took, a sound she associated with antique stores back home. The shop was filled with many household items: baskets, light and dark, large and small, simple and fancy; shirts, hats, dresses, and shoes; toys; and gifts. Erin was drawn to the assortment of moccasins displayed against one wall. She fingered the fine leather.

"I've always wanted a pair of moccasins," she said wistfully.

"You should get them."

Erin looked at the prices of the style she liked. "If I bought these," she said, indicating the white leather moccasins, "I wouldn't be able to afford anything for anyone else back home."

Neyse thought a moment. "You can save the money when you get home, send it to me when you have enough, and I'll come get them for you."

Erin brightened. "Would you?"

"Of course. Try them on so I'll know what size to get you."

Erin tried on various styles and sizes and decided, at last, that a size eight was the best fit. Erin rubbed the soft leather before putting them back into the box.

"I'm going to need a part time job when I get home," she said as they left the store.

"You might be able to find something on campus. That's where I'm going to look."

"Just think…a month from now we'll both be in college."

Neyse nodded. "I'm excited and nervous at the same time."

"I feel exactly the same way!"

They saw Julia right away, sitting on a bench near the gazebo.

"Having fun?" she asked as they approached her.

"Yes!" both girls said.

"Hungry?"

Neyse nodded, nudging Erin to agree.

"Why don't we have Mexican food?" Julia suggested. "They have some of the best restaurants in the state here."

They walked around the square, read several menus, and finally made their decision. As they entered the restaurant, Erin noticed the uneven brick floor, which looked very old and worn. The bricks held their temperature and helped cool the building, providing instant relief from the heat outside. Julia gave her name to the hostess and they were greeted and taken to their table.

A brightly clad waitress brought them water, menus, and chips with salsa. As they considered the luncheon selections, all three of them dipped their hands into the basket of chips. By the time they were ready to order, the basket was empty.

"We'd better quit eating chips," Julia said after the waitress had taken their order, "or we'll be full long before our tostadas get here."

Erin put her napkin over the basket.

"So, what's in the bags?" Julia asked.

The girls told her about their purchases.

"I want to get something for my dad," Erin told her

aunt, "but I don't know what."

"He always likes books," Julia said.

"You're right, but which book is the question."

"I need to go to the bookstore at the university after we're done here. I'll help you find something for him."

"Thanks, Auntie."

After they finished lunch, Erin took out a twenty-dollar bill and handed it to Julia.

"What's this for?" Julia asked.

"I owed Neyse this lunch," she explained. "I lost to her at pool."

"Put your money away. This is my treat."

"But…"

"Put it away," Julia said.

They headed across town to the University of New Mexico, through streets crowded with cars, every one in a hurry. The heat seemed to make drivers impatient. People honked, crowded intersections, and had no patience for pedestrians.

"Mom! Look out!" Neyse cried.

Julia slammed on her breaks, barely missing a driver who had merged into their lane without looking.

"I hate this traffic!" Julia said.

"I'm glad you're driving and not me," Neyse said.

Erin felt as if they'd stepped into an accelerated movie, the images jerky and unnatural. She'd become so accustomed to the slow pace on the pueblo that the bustle of the city made her anxious. She was relieved when they finally arrived at the campus.

Julia parked on a tree-lined street where the shade lowered both the temperature and the tension.

The entire campus was built in the pueblo style, the simple lines of the buildings both familiar and appealing to Erin. She was surprised at how serene and comfortable the architecture made her feel.

They only saw a handful of students as they followed

Julia to the bookstore. She found the volumes she needed and a book on art of the southwest for Erin's father. Neyse wanted to get a sweatshirt, but couldn't find one in a color she liked in her size.

On their way back to the car, Erin took off her shoes and walked barefoot on the well-tended, green grass. "I'm always surprised at how being hot can make you tense and being cool can relax you."

Julia laughed. "Unless you're in front of a fireplace at a ski lodge."

Erin laughed, too. "Yeah, unless that."

By the time they reached the pueblo, the wind was blowing up another storm. They drove by the canal and Erin saw a strange sight. Wispy white clouds like feathery snowflakes, floated in the air by the water.

"What's that?" Erin asked, pointing.

"That's from the cottonwood trees," Julia told her. "When the pods break open in the fall, it looks just like it's snowing." She pulled to the side of the road to give Erin a better look.

"It *does* look like snow," Erin agreed. She took out her phone. "Can I take a picture?"

"Of course," Julia said. She smiled and nodded. "Winter will come early this year from the look of it."

Erin was surprised to feel the sting of tears in her eyes. Winter. The summer was ending. She'd have to go home soon; leave the pueblo and Neyse. And Paolo.

Julia pulled back on the road. "Have you two decided what you want to do for your birthdays?"

Erin was glad that Neyse answered for she wasn't sure that she could trust her voice.

"Juan and Paolo are planning a surprise for us."

"What kind of surprise?"

"They haven't told us."

"When's this going to be?"

"The Saturday after our birthdays."

"In that case, we can celebrate the family night on your birthday."

"Which one?" the girls said in chorus.

"Neyse's. Erin will have the advantage of an early celebration. I want you to get together and decide what you want for dinner."

"Our choice?" Neyse said.

"The same as always, anything you want."

"Oh boy!" Neyse said happily. "This year is definitely going to be better than the average birthday.

"When did you ever have an *average* birthday?" her mother asked.

That evening, as Neyse set the dinner table, Julia asked Erin to run to the field for squash.

"Sure, Auntie."

Walking down the hard-packed dirt road, Erin felt the quiet peacefulness of the pueblo. Albuquerque had been so hyper and noisy and harsh. She had gotten so tense with the commotion. Here she was relaxed and calm. She smiled to herself. *It's nice to be back home again.*

How easily she'd adapted to the relaxed village pace.

How strange the outside world had become.

CHAPTER 17

"I'm exhausted," Neyse moaned, pushing the Saturday morning paper off the chair and plopping down next to the ice cream maker. "I can't turn that crank even one more time."

Juan stepped grandly up to the machine. "It's a good thing that the *men* are here or we wouldn't be having homemade peach ice cream for dessert tonight."

Neyse jumped up and pushed him aside. "Forget it, I'll manage. Come and help me, Erin."

Erin started to get up from her seat on the couch but Paolo grabbed her arm. "Don't spoil the fun. Let us rescue you girls."

"Oh, nuts," Neyse said, but she sat down again none-theless.

Juan began to crank the ancient machine, clearly a relic from days long past. "We probably won't have much appetite for this after Indian tacos and beans, anyway."

"Just don't make a pig of yourself and you'll be fine," Neyse told him.

Julia had already begun cooking the fry bread and the tempting smell drifted into the living room.

"Do you need any help, Auntie?" Erin asked.

"Not tonight, thanks. The birthday children don't have to help in the kitchen."

Tony and Carlos were busy preparing vegetables and setting the table. The front door opened suddenly and in streamed Victoria, Ramon, Felicia, Alonzo and Luis.

"Happy birthday," they all cried out, one after another, each greeting Neyse and Erin.

"Where should I put these presents?" Victoria asked.

Neyse took the brightly wrapped boxes from her and put them with a stack of others already on the corner lamp table. Victoria went into the kitchen and began to help Julia carry serving dishes to the table.

"Is the ice cream finished?" Tony asked.

"I don't think so," Juan said. "Shall we open it and check?"

"*Ha'a*," Tony said.

They all gathered around as the machine was unlatched and the dasher pulled out. Tony ran his finger along one blade and tasted it.

"About ten more minutes, I'd say. It tastes wonderful," he added with a smile at Neyse. "Good choice."

Juan put the machine back together and Paolo took hold of the crank. "I'll finish it."

"But dinner's ready," Julia complained.

"No problem," Neyse said. "We'll feed him while he's cranking."

"We need an electric ice cream maker," Julia grumbled loudly as she put a bowl of steaming beans on the table.

Erin noticed Tony's look of sharp disapproval. He started to say something but checked himself. She wondered if her aunt and uncle ever really argued the way her parents did. She'd never heard them. Maybe Indian wives weren't supposed to argue with their husbands.

"Please, everyone, sit down," Julia urged.

The family crowded around the table and passed the

food back and forth until each plate was loaded almost to overflowing.

Tony reached around Victoria and poked his youngest niece in the ribs. "Felicia, that taco is bigger than you are."

"It is not," she argued.

"Bet you can't finish it," he said.

"Bet I can." She took an imposing bite.

Neyse offered to feed Paolo but he refused. "Ice cream's almost done," he told her.

A few minutes later everyone agreed that the ice cream was finished. Paolo put the canister in the freezer and squeezed in next to Erin at the table. She loved to have him sit so close to her.

"You haven't eaten much," he observed, looking at her plate full of food.

"I was waiting for you."

He prepared a bulging taco for himself and took an outrageously big bite.

"I can see where Felicia gets her ideas," Tony said.

Paolo winked at his uncle, nodding.

"Have you heard from your parents, Erin?" Victoria asked.

Erin nodded, quickly chewing and swallowing what she was eating. "Yes, they called on Sunday and I got a package from them today."

"I'll bet they've really missed you this summer," Victoria added, more to herself than to anyone else.

"Will you open the presents before or after ice cream?" Felicia asked, looking mournfully at what remained of her taco.

"Before," Neyse told her.

"Good," came the relieved reply.

Julia and Victoria did the dishes and soon everyone was seated around the living room.

"If you'd hurry up," Carlos said, "Luis and I could ride our bikes for a while before it gets dark."

"How rude," Neyse said.

"I, for one, am ready to hurry up," Erin said with a nod to Carlos.

"You have to wait," Paolo told her. "It isn't your birthday until tomorrow."

"That's not fair," she said. "The celebration is today."

"Right," Julia said, sorting the gifts and giving each girl a stack of presents.

"How shall we do it?" Neyse asked, looking at hers with interest.

"What about taking turns? You open one, and then I'll open one," Erin suggested, as she poked curiously at one package.

"Perfect," Neyse said, tearing open the first card before Erin could say another thing. "This is from Mom and Dad," she said. The gift was in a small, rectangular box. She opened it and took out a piece of paper and a note. "It's a gift certificate," she squealed, "for the pet shop in town. For a puppy, right?"

"Right," Tony said.

"Really?"

Her mother nodded.

"Any kind I want?" Neyse popped up and hugged them both. "Thank you," she bubbled.

"You're welcome, dear," Julia said.

"Can we get it this week?"

"I think it would be better to wait until after Erin leaves. A new puppy will take all your attention for a while. That isn't fair to Erin on her last week."

Neyse nodded her understanding. "You're right. I'll wait." She turned to her cousin. "Your turn, Erin," Neyse said, her excitement just below boiling point.

Erin had looked at the assortment of boxes and cards in front of her. She knew which were from home, from Neyse, from Julia and Tony and from Victoria. But there wasn't anything in the stack from Paolo. She was disappointed but did her best not to show it.

She chose the card with Neyse's handwriting on it. Juan had also signed the greeting inside. Erin opened the brightly wrapped box and found a coordinated selection of writing paper in six different shades of pink, lavender and purple with matching envelopes and pens.

"Oh, Neys, they're super. Thank you."

"I thought this would encourage you to write me after you go back home. Real letters, not email, right?"

"I will, honest. I promise."

Erin hugged Neyse, feeling happy and sad at the same time: lucky to have a cousin who was so special and sad because she would not see her for at least another year.

"Open ours!" Felicia cried out to Neyse.

"Okay," she told the excited child with a smile.

The box from Victoria was heavy. Neyse frowned slightly, obviously having no clue what it was as she lifted it from the floor onto her lap. She pulled off the ribbon, tore the paper and opened the lid. Inside it, in another box, was an iPod.

"My gosh," Neyse exclaimed. "This is terrific. I've wanted one for ages."

Victoria smiled. "So we heard."

"Now you open ours, Erin," Felicia encouraged.

"All right."

Erin's box was light and lovingly wrapped. In it, nested in a fluff of pale pink tissue, was a beautiful lavender angora sweater. She held it up and smiled at her second aunt.

"It's gorgeous!" she cried. "I love it."

"I hope it fits," Victoria said.

"I'll wear it whether it fits or not," Erin promised.

Neyse opened Erin's gift, a basket of bath salts, body lotion, potpourri, and nail polish that she'd found in a specialty shop Neyse had liked when she had visited.

Julia and Tony gave Erin a wool skirt to match the sweater Victoria gave her. She thanked them several times, feeling uncomfortable, knowing how expensive the brand was. She tried to push the discomfort away.

Juan's gift to Neyse was a turquoise and silver bracelet, made by a friend in northern New Mexico. Once again, Erin felt the disappointment that there was nothing from Paolo.

After the last gift had been opened Carlos and Luis jumped up.

"Don't you want any ice cream?" Neyse asked.

"Later," they said, and ducked out the front door.

"We'll dish it up," Juan said, nudging Paolo.

As he stood, Paolo took a small card out of his shirt pocket and handed it to Erin. On the envelope was printed *HENATEETZ*. The short note inside said: "Your gift has been ordered but won't be ready until next week. Happy Birthday, my Cloud Dancer. Yours—Paolo."

Mine, she thought happily. *Mine*.

"What does it say?" Neyse asked.

"It says that my present from him won't be ready until next week."

"I wonder why?" Neyse asked, looking innocent.

After a considerable time and a good deal of racket, Juan appeared with a large tray holding several bowls of ice cream. Two of the bowls sported candles sputtering just above the ice cream.

Erin and Neyse made silent wishes and blew them out, as everyone sang "Happy Birthday." The bowls were passed quickly, as the ice cream had already begun to melt.

"This is delicious," Victoria said. "Is there enough for seconds?"

"There was hardly enough for firsts." Juan told her.

"That's what happens when you put the men in charge of the food," Neyse said.

"You're asking for it," Paolo warned her.

Alonzo left after eating his dessert. Everyone else stayed to play charades. The game ended just after eleven o'clock when Felicia fell asleep on Ramon's lap.

"I guess we'd better go," Victoria said.

After they'd left, Julia started the dishwasher while Tony emptied the trash. "We'll see you in the morning," Julia said.

"Yes, goodnight," Tony added.

"Thank you for the party," Erin told them.

"You're very welcome, Erin," Tony said.

Julia agreed with a nod but her tired eyes said she was glad it was over.

Neyse turned off the radio and all but one light in the kitchen. She sat down next to Juan and snuggled up against him. Erin wondered if Paolo would feel uncomfortable and leave. He didn't. He sat down on the couch, put his arm around her and pulled her close to him.

Carlos chose that romantic moment to sneak into the house, bringing laughter from the four friends.

"Do you guys have Saturday all planned?" Neyse asked.

"We're not talking," Juan said mysteriously.

"You have to tell us something. We need to know what to wear and when to be ready."

"We'll get back to you," Paolo told her.

"Come on, you two," Neyse whined.

"It's a surprise. You'll have to wait," Juan insisted.

"Give me a hint," Erin whispered to Paolo.

He smiled, conspiratorially. "No," he whispered back.

She started to pull away but he held her tightly and whispered to her, "Be tough. Be like me."

It sounded like teasing though she suspected it wasn't entirely that. She started to argue that she wasn't like him, she was an outsider. The fact that he'd suggested she could act like him stopped her. She turned her head slightly, smiled and nodded. Then she wondered if an Indian would have bothered with the smile.

She giggled. "It's not easy to be an Indian."

"That's what I've been telling you."

They laughed, enjoying the closeness.

"What's so funny over there?" Juan asked.

"Be tough," Paolo told him. "Don't ask."

Juan gave him a puzzled look and shrugged his shoulders. "Private joke, I guess," he said to Neyse.

Paolo kept looking at the clock and Erin was mystified.

"Do you have another date?" she asked.

"Would you be upset if I did?"

"*Ha'a.*"

He squeezed her. "I don't. There's just something I need to do tonight."

"What?"

"You'll be the first to know."

Juan and Neyse were lying on the floor now, talking quietly and kissing. Erin felt completely relaxed being with Paolo in the same room with them, confident that he wouldn't get carried away. She felt her own excitement as he kissed her, coaxing her with his lips until she finally let his tongue dart into her mouth. Her stomach flip-flopped and she shuddered, tingling in every part of her body. Her heart pounded and she felt almost as if it might explode.

He craned his head around and looked at the clock again. "It's almost midnight," he whispered. "I want to be the first to tell you: Happy birthday."

She smiled and kissed him, testing to see if he wanted the touch of her tongue. A surge of electricity passed between them. "Let's go for a walk," he said quietly.

"Now?"

"Yes."

"I don't think we should."

"Why not?"

"It's too late."

"Are you still afraid of me?"

"No, I just don't think we should go out this late."

He stood abruptly and went into the kitchen, where Erin heard him get a glass of water. She went after him.

"Maybe if Neyse and Juan weren't here. It just doesn't

seem right this way."

"Never mind."

"Please don't get mad, Paolo. I want to be with you. I'd rather be alone. Now just doesn't seem right."

She put her arms around his neck.

She looked up at him, knowing he was fighting with himself.

"I'd better go."

The tone of his voice said his mind was made up. She followed him as he walked quickly to the front door.

Juan looked up. "Time to go?"

"I think so," Paolo replied, his voice strained.

Paolo led Erin outside and put his arms around her. "Happy birthday, *Henateetz*....my Cloud Dancer."

"My happiest ever," she whispered, contentedly.

Erin and Neyse picked up the mail on Friday afternoon and found a card addressed to both of them. It read: "Be ready for your surprise at nine o'clock tomorrow morning. Dress casually. Bring a camera."

"What in the world do you suppose they have planned?" Neyse wondered aloud.

Erin shook her head. "I can't imagine."

That evening Neyse decided to take a long, hot bath, and use some of her new bath salts.

"Take your time," Erin told her. "I'll shower in the morning."

The house was quiet, for Tony and Carlos were out fishing with Alonzo and Luis. Erin went into the living room just as Julia turned on the vacuum. She went over and over the rug in front of the door, a frown on her face. Finally she turned it off and slumped down in a chair.

"I get so tired of the dirt! I can never seem to get the house really clean. And when it's windy, grime is everywhere."

Erin felt as if she were hearing a conversation not really

meant for her. "Do you want me to dust, Auntie?"

"Oh, Erin, I wasn't asking for help. I just get tired sometimes, that's all."

"Tired of the dirt?"

"Tired of the struggle. I'm fighting a battle I can't win. I can't make the pueblo what it isn't. It's dirty here. There are no paved streets, no lawns, no flower gardens, no filtered air systems." She said the last part with a laugh. "The people here don't care. It has no importance to them. They never had any of those things so they don't miss them. Here they cherish the spirit, the soul, the living of life. That's where the value and meaning is for them." She stared out the front window.

Erin felt awful for her.

"Would Uncle Tony move?"

Julia shook her head. "Never. This is his home—physically and spiritually. I agreed when we were married that I'd live here. I wouldn't ever ask him to leave. Sometimes I just get tired. That's all."

"You're not sorry you married him, are you?" she asked, surprised at her own boldness.

Julia thought for only a moment and then smiled. "No. If I'd married an Anglo I'd probably complain that he spent too much time at the office or he didn't mow the lawn often enough. No matter who you marry, there are always compromises. Tony makes them just as I do."

"It seems to me that you have had to give up more than Uncle Tony has," Erin said, her voice soft, not wanting to hurt her aunt's feelings.

"I thought his culture had more to offer me than mine had to offer him. I respect the pueblo way of life more than I respect the way of the outside world. It's slow moving here. We aren't always pushing or striving or going after some new thing. Indians simply are. They live. They share. The struggle here is not to go somewhere or gain something, but simply to be."

"You sounded like you weren't very happy a couple of minutes ago."

"Is your mother always happy?"

"No, of course not."

"Nobody is always happy, sweetheart, unless they are lying to themselves."

Julia stood, kissed Erin on the cheek, and turned the vacuum back on.

That night after she went to bed, Erin thought a long time about what Julia had given up to live on the pueblo. She wondered if she'd be able to do the same.

They were ready at nine. Erin wore a pair of jeans and a pale blue cotton shirt; Neyse, black pants and a red pullover. Erin put a freshly charged battery in her camera. It seemed like forever before they heard a motor in the driveway and then a knock at the door.

Neyse answered it. "Gee, you guys look nice."

They both had on corduroy pants. Paolo wore a brown plaid shirt, Juan a cream colored pullover.

"This old thing?" Juan teased.

"I gave you that 'old thing'."

Erin looked at her watch and frowned at Paolo. "It's about time you got here."

"Haven't you learned yet? It's not 'about time', it's Indian time."

Juan laughed but Erin pouted.

"Come on," Paolo said, "don't be grumpy. We're always late here." Then he added more quietly, "You'd better get used to it."

Erin couldn't help being frustrated. It made her feel that she wasn't important enough for him to make the effort.

They piled into the car and Juan started the engine. Tony and Julia waved as the foursome pulled out of the driveway and headed toward the main road.

"Have fun," Julia shouted.

Juan took the highway toward Albuquerque, merging easily into the Saturday morning flow of traffic. Neyse and Erin looked quizzically at each other, for no information had been forthcoming as to the destination of their journey. They turned off the highway, finally, onto a road where a huge sign said Sandia Peak.

Neyse smiled in recognition.

"What is Sandia Peak?" Erin asked.

"You'll see," Paolo told her.

Before long Juan pulled into the parking lot and found a space where he opened the door and hopped out. Paolo reached into the trunk and brought out a picnic basket.

"What is this place?" Erin insisted.

"Patience," Paolo whispered.

They walked across the parking lot toward a building where people stood waiting in a line. It was then that Erin saw the bright red tram, inching its way up the steep mountainside. She flashed a smile at Paolo.

"We're going in that?"

"Part of our research."

She laughed.

Juan and Paolo bought tickets and they boarded the second tram which had just returned down the mountainside. They crowded against a window so that they could see the spectacular view.

"Have you ever been here before?" Erin asked Neyse.

"Not since she was little," Juan answered for her.

Neyse look confused. "How did you know?"

"'Cause we asked your folks."

They rode higher and higher, seeing an increasingly breathtaking view of Albuquerque and the Rio Grande Valley far below. Erin was a little nervous, but the ride was smooth and she was fascinated by the scene.

"Aren't you going to take any pictures?" Paolo asked.

"Oh, yes!"

She took her camera out and snapped several shots both

of the valley and of the mountains. Once at the top, they walked along the observation deck. Erin took pictures of Juan and Neyse as they stood by the guard rail, the Rio Grande Valley behind them. The sky was a pale blue, and filled with feathery white clouds. A slight haze in the valley below gave the view a muddy gray cast. Although the sun was hot, a chilly wind blew across the mountain, and Erin shivered.

"Are you cold?" Paolo said, putting his arm around her.

"Not now," she said, happily.

"What did you think of the tram?"

"It was terrific."

"Is it the kind we should get for the ruins?"

"I think it's a little large for our needs."

"I agree. Maybe a chair lift would be better, or an Escalator."

"We could get a used wheelchair and a rope," she said with a grin.

He shook her gently. "You're awful."

They checked out the restaurant and the ski lifts, and took pictures of each other clowning.

"Is anybody hungry?" Paolo asked. "I'm getting tired of carrying this basket around."

"Be tough," Erin teased. "Don't complain."

"I'm hungry," Juan said.

"Me too," Neyse agreed.

"Let's get away from all these people," Juan said quietly, leading the others up a hillside to a remote spot.

"It sure looks good," Neyse said as Paolo took the food out of the basket. He'd brought hunks of cheddar cheese, ham, huge green apples, cupcakes, and four canned drinks wrapped in aluminum foil.

"Mom made this just for you," Paolo told the girls as he unwrapped a golden loaf of bread. "We wanted to have cold whale filets but they wouldn't fit in the basket."

Erin took several pictures of the lunch and even asked a

passerby to take one of the four of them together.

"Let me get one of you two," Neyse said, reaching for the camera.

Erin and Paolo posed, made faces, and tried looking serious but they could not. Finally, they looked at each other and Neyse snapped the best photo of the day.

Juan and Paolo ate quickly but the girls lingered, taking time to enjoy the food, the view, and each other.

When they'd finished they sat contentedly looking at the magnificent expanse of land. In the valley to the east lay the huge city of Albuquerque; to the north, beyond view, was the pueblo. Two worlds, so far apart, Erin thought. *How can they come together?* She didn't want to consider the obvious answer.

"Enough sitting," Juan said.

They walked back toward the restaurant, taking time to look at the scenic post cards offered for sale, several of which showed the mountain in winter, covered with snow.

"Have you ever skied here?" Erin asked Paolo.

"Skiing is for rich people," he replied.

She stiffened as she felt the wall come between them immediately.

"I'm sorry," he said. "No, I've never skied at all. Have you?"

"Lots of times. I'm not very good at it, but I love it."

"I think I would, too," he admitted and the wall crumbled. "Why don't you come back during winter vacation and teach me?"

"I wish I could. I doubt that my folks would want me to leave at Christmas."

"Do you think you'll come back next summer?" he asked in a low voice.

It hurt, but she said it. "I don't know," and she felt the warmth of the summer slipping away.

CHAPTER 18

Hot days alternated with cool nights the following week, Erin's last week on the pueblo. Erin, Neyse, Juan and Paolo spent every possible moment together, at the rec center when Paolo was working, at the lake when he wasn't. They tried to squeeze all they could into the short time they had left. Erin's plane was scheduled to depart from Albuquerque Airport on Sunday morning at eleven.

As the week progressed, Erin wondered more and more about the gift from Paolo. He hadn't mentioned it again since the night of the party and she hadn't been able to get any clues from Neyse. Her curiosity picked at her, teasing her with thoughts of possible presents. Perhaps a book or a CD. *What else could he have ordered?*

She'd thought for sure that whatever it was would have arrived by Friday. But the foursome had spent the entire day at the lake and nothing had been mentioned about it. Juan and Neyse had gone for a walk along the beach while Erin and Paolo enjoyed the warm sun on the sand.

"I have to go to my cousin's tomorrow," Paolo said, "to pick up some things. Would you like to go with me?"

His voice startled her. He'd been quiet for so long that

she'd thought he was asleep.

"Tomorrow is my last day here."

"I know. I'd like to spend it with you."

She was delighted to hear him say it but not sure what her aunt's response would be.

"I'd have to ask Aunt Julia to see if she'd mind."

"I already asked her. She said it would be fine."

"Oh. Well then, yes. I'd like to go."

How strange. Why would he ask Aunt Julia if I could go with him to his cousin's?

Juan and Neyse came back from the water's edge, Neyse wet and shivering. "The lake's starting to get cold."

"Oh, Neys," Juan said.

"It is," she said through chattering teeth.

"To hear you tell it, we'll be ice skating on it by Halloween."

They all enjoyed the joke, but for Erin, the laughter at the summer's end could not match the carefree sound it had had at the beginning.

"What's on for tomorrow?" Neyse asked. She dried herself with a bright orange towel and pulled on a shirt.

Paolo answered her. "Erin and I are going down to my cousin's."

Erin expected Neyse to object to her leaving on her final day. Instead, she flashed a big smile at Paolo and said, "Good."

"Do you want me to fix a lunch?" Erin asked him, feeling more relaxed about the whole idea. Neyse's approval made it much easier.

"No, we'll stop somewhere."

The next morning it was hotter than it had been all week.

"Do you think I should dress up?" Erin asked Neyse as she looked through the drawers.

"I wouldn't," her cousin replied.

"But I've never been to his cousin's before."

"You can be sure that James won't dress up," Neyse said with a grin.

Erin shrugged. "*Hin'a,*" she said, easily using the now familiar word.

She put on her favorite, the white shorts and a white cotton blouse, and then braided her hair. The sun had lightened the top layer so much that it now looked as if it were braided with strands of gold.

Paolo arrived shortly after she'd finished eating breakfast. He spoke with Tony for a few moments before coming into the house.

"Are you ready?"

She nodded.

Neyse hugged her. "Have fun."

"We will."

"You'll be back for dinner, right?"

"Yes," Paolo told her.

"And you're staying?" she asked him.

"Would I turn down your mom's cooking?"

Paolo put his arm around Erin as they walked outside. He'd dressed casually, too. She loved the way he looked in cut-offs and a plaid shirt, the sleeves rolled up, his arms trim and tan.

"Tiya at home?" she asked as he opened the truck door.

"Yes. This day is just for you and me."

She gave him a special smile. "I'll bet Tiya is glad I'm leaving tomorrow."

His jaw clamped tightly. "But I'm not."

Erin sat next to him in the cab of the truck, happy to be with him and at the same time sad, knowing it was their last day together. He said nothing as he drove but the silence didn't bother her. Erin knew that it was close to ten o'clock when they left home and almost eleven when they arrived at his cousin's, yet it seemed that the trip had taken no time at all. Today there would be no time; time sat waiting at the edge of this day.

James lived on a pueblo much like their own. Paolo drove by a *Kiva* on the way through the village and Erin remembered how curious she'd been when she had first arrived. She didn't wonder about the *Kiva* anymore. Strangely, it didn't matter to her. It had nothing to do with her. She finally understood what Julia had said.

James greeted them and invited them into the house.

"Hello, Auntie," Paolo said to a woman in the kitchen, who came to meet him as soon as she heard his voice.

"It's been too long," she said, more as a welcome than a criticism.

"Yes," he agreed. "Auntie, this is Erin, Neyse's cousin."

Erin thought she detected a hint of disapproval in the woman's glance but she smiled and said hello.

"Please sit down," she said, offering Erin a chair.

Paolo and James left the room without further word.

"How long have you been visiting?"

Erin noticed right away how much older she looked than either Julia or Victoria. Her face was lined and her hair was already mostly gray. She wore an old woman's dress of a drab green. Her dark face had no make-up.

"All summer," Erin answered.

"Do you like it here?"

"Yes, very much."

Erin was sure that the woman wasn't pleased with that answer.

The conversation continued, formal and slightly strained. She'd never felt more like she didn't belong, almost as if she had *Outsider* printed on her forehead. She could think of little to say to this unwelcoming stranger.

James and Paolo came back into the room. She jumped up, relieved.

"I see you're busy, Auntie, so we won't stay."

She made no pretense of arguing with him.

"I just made some prune pie. Wait a minute and you can take some with you."

She wrapped several pieces of the still-warm, sweet pie and put them in a bag. Paolo thanked her and they left.

Erin hadn't seen Paolo get anything from James. Still, she knew better than to ask about "things" so she kept her questions to herself.

They drove for a while until Paolo found a spot along the river he seemed to be looking for. He pulled the truck off the road, parked, and got a coarse, brown blanket out from behind the seat.

"Let's go sit by the water."

They walked down to the river, where Paolo spread the blanket under the cottonwoods. They sat down next to each other. Erin leaned up against him, feeling utterly content. The river danced and sparkled in the midday sun.

Neither spoke for a long time.

Erin turned finally and looked at him. "Remember the first day? You took me to see the river then, too."

"The canal."

"It seems like so long ago and yet it went by so fast. I can't believe I'm going home already."

He was quiet for a while and then said, "Are you excited about college?"

She nodded. She waited for him to say something else but the silence lingered. After a long time she decided broke the silence. "Why aren't you going to school, Paolo?"

He shrugged. "What for?"

"For an education, of course."

"So I can join the rat race?"

"No. So you can do more to help the pueblo. That's what you want, isn't it?"

"How can that help?"

She shook her head. "Maybe by learning better farming practices; maybe by representing the pueblo people in state politics; maybe by getting grants to improve housing or health care for the Indians." And then she added gently, "I think it might be more satisfying than just helping Mr.

Montoya with his field."

She knew he might get mad at that but she said it anyway.

She waited.

After a while he nodded and said, "I'll think about it."

"Let's go in the water," she suggested, hoping to lighten their mood.

"*Hin'a.*"

They took off their shoes and walked down to the water's edge.

Erin stuck her foot in the rippling flow. "It's cold."

Paolo walked in up to his knees and grinned at her. "Come on, *Saucha,* it isn't that bad."

"*Kotra auksi,*" she said, making a face at him.

He looked at her with pride. "You're learning."

They walked upstream, stopping on partially submerged rocks to watch the small, darting fish.

"'Beauty below me…'" Erin said.

"What?"

"It's from a poem Neyse told me about. I think it's Navajo. I can't remember all of it but it says that there is beauty everywhere. I admire the way Indians see the beauty of creation."

"A friend of mine told me that you don't have to be Indian to appreciate beauty."

"Oh, really?"

"Yes. The same friend says, 'Floss your teeth every day.'"

She gave him a shove, a little harder than she intended, and he went over into the water.

"Oh, Paolo!" she gasped, but when she saw the look on his face she knew an apology would not be enough.

She started to run, but be grabbed her.

"I've already been baptized," she shouted, to no avail.

He dunked her.

She came up sputtering. "*Ma'a!*" she yelled.

He grinned at her. “You started it.”

“And you certainly finished it,” she said, wiping water from her face.

He took her hand and led her back downstream, where they moved the blanket into the blazing sun and stretched out to dry.

“Are you hungry?” he asked.

“Some.”

“We can go get something to eat.”

She thought for a moment.

“You know what I’d really like?” she asked.

“What?”

“Prune pie.”

His eyes twinkled. “You’re turning into a pueblo girl.”

Pleasure filled her as she watched him walk to the truck and bring back the bag his aunt had given him. At the same time, he carried a small bundle that he put on a corner of the blanket. She was curious about the oddly shaped parcel but she did not ask about it.

Paolo unwrapped the pie and gave her a piece. She ate a bite and then offered him some.

“How is it?”

“Almost as good as Aunt Julia’s, whose is almost as good as your mom’s.”

“You’re a regular diplomat.”

They ate four pieces between them, Paolo mentioning that he wished they had milk to go with it.

“As I recall, you were in charge of lunch,” she teased.

He raised his eyebrows. “Do you want to go swimming again?” he asked, threateningly.

“No. I’m just saying.”

After they ate, Paolo placed the mysterious package in front of her. It looked like two wadded up bath towels.

“Happy birthday, Cloud Dancer.”

“*Nachra.*”

He nodded.

Slowly she opened the rough, fawn-colored cloth, unfolding the gift hidden within.

Her eyes widened. She looked quickly up at Paolo and then back to her treasure: a pair of cream-colored leather moccasins, similar to the ones she'd seen in Old Town, but softer and finer.

She looked at him again in amazement. "Where did you get these?"

"James made them for me, for you."

"But, how did you know?"

"Neyse told me. I'm glad that you didn't have enough money to buy the ones in town."

She brushed off her feet and carefully slid them into the soft leather moccasins, tying the thin strips of rawhide on each side.

"Paolo, they're perfect. How did you know my size?"

"Neyse, of course."

"I don't know what to say. They're beautiful."

"Do you like them?"

"Oh, yes! They're wonderful."

"I wanted you to have something special so you wouldn't forget this summer."

She looked at him and frowned. "I could never forget, Paolo. Never."

He was quiet for a long time.

"Erin, there are things I'd like to say to you, but I can't. I can't because you're an outsider and because you're going away. You're going back to your life and I'm staying here. I'd like it to be different; I would like you to stay, but you can't."

The tears spilled down onto her cheeks.

"Don't cry, Little Cloud."

He put his arms around her. "We're different, Erin. We're different in so many ways. It may not seem like it to you but it's true. This summer has been like a dream, but every dream has an end."

"But if I come back next summer..."

"Don't live for next summer. Live for now. You'll meet others this year. If you allow yourself, you'll find someone you like. Don't put me between you and another."

In a whisper she asked, "Will you meet others, too?"

He looked away, toward the river. He said nothing. She waited but he was quiet. She leaned toward him, impatiently, and saw the tears in his eyes.

He wiped them away quickly and faced her again. "Yes, I will meet others."

She felt the coldness she'd seen the first day. She put her face in her hands. She couldn't pretend as he pretended.

Gently he pulled her hands away but she refused to look at him. He leaned down and kissed her.

"*Henateetz...*" he whispered. "*Henateetz.*"

He drew her down on the blanket and held her until her tears had dried. He kissed her, holding her so close to him that it hurt.

"What if I come back next summer, Paolo? Tell me the truth."

Again he was quiet.

When he looked at her, his face was soft. "I will still want you," he whispered. "I'll always want you."

She reached for him and they kissed, their hearts, tender and exposed, not too young for breaking.

She ached for what she knew she could not have.

Please. Please say it. Tell me you love me. I've never heard it before. I want you to be the first.

But he'd already told her there were things he could not say and they both knew that was one of them.

Time swept past them like the river, flowing silently, not touching them at all. They held each other, wondering if there would be another summer.

Their clothes were long dry, the sun far down in the sky. "We'd better go," Paolo said hoarsely. "I don't want to upset Aunt Julia on your last night."

Erin sat slowly and drew up her legs. She began to untie the laces on the moccasins.

"What are you doing?"

"Taking them off."

"Why?"

"I don't want to get them dirty."

"Outsider," he grumbled.

She put them back on, waited for him to fold the blanket and then walked defiantly up the path to the truck.

He smiled.

Once in the truck, she carefully dusted them off.

"It isn't easy to live in another culture," he observed.

"I know." And she did.

Neyse practically did cartwheels when she saw the moccasins.

"Aren't they beautiful, Erin? Don't you love them?"

Tony and Julia admired them as well, the craftsmanship clearly exceptional.

During dinner Neyse kept asking, "Were you really surprised? Did you have any idea at all?"

Finally Erin said, "I thought he'd forgotten my birthday altogether."

His look said he'd never forget her or her birthday.

Before he left, they shared a long, private goodnight. She sat in the back of the truck with him, watching the stars emerge in the darkening sky. They didn't talk, since all there was to be said, had been said that afternoon.

It was very late when Erin began packing. She sat for a long time holding the moccasins.

"I'd like to wear them but I don't want to ruin them. I want to keep them...forever."

"They wouldn't look very good with the outfit you're wearing anyway," Neyse told her.

"Well, I'm not putting them in my suitcase. Luggage gets lost all the time. I'm not going to take any chances. I'll take them in my backpack."

She wrapped them carefully in a blue tee shirt and put them on top of everything else she'd already put in her backpack. As she zipped it up she remembered packing it in California weeks before. She thought of her camera, her arrival, of what she'd brought to the pueblo and what she would take home. The summer had been full of many gifts.

The airport was crowded and noisy, a nasal voice on the loudspeaker announcing arrivals and departures. Tony and Paolo carried Erin's suitcases up to the counter and checked them in. Neyse was holding onto her backpack.

"Your flight will leave from gate six," she was told.

Paolo held her hand as they walked along the concourse to the security area. Juan and Neyse struggled to walk beside them but the oncoming flow of incoming passengers made it virtually impossible. "I hope you managed to get everything packed," Julia said once they'd gotten to security.

"Don't forget, we want you to come back next summer," Carlos told her.

She squeezed Paolo's hand. "I won't forget."

"You're welcome anytime," Tony said.

She nodded. "I know that. You've made me feel it from the first day."

"We are really going to miss you," he said tenderly.

Too tenderly.

She began to cry. She hadn't wanted to. She had told herself that she would be strong, but her tears won the battle.

As Tony hugged her, her flight was called.

"Goodbye, sweetheart," Julia said, kissing her. "Please come back soon."

"I will. I promise."

Carlos hugged her and then Juan. "I hope you have a whale of a good year," he teased.

She laughed. "You've been saving that one, haven't you?"

She and Neyse embraced. "You're my sister now," Neyse said. "The room is half yours, so don't stay away too long." She handed Erin the backpack.

Erin wondered what Paolo would do in front of her family. Everyone moved back away from the two of them, giving them a small, private space.

He put his arms around her, looked at her tenderly, and then kissed her.

"Don't cry, Little Cloud," he said very quietly. "We'll all be here waiting for you when you come back. *All* of us," he said, and then added softly, "especially me."

Then, lovingly, he kissed away her tears.

Erin heard her flight called.

She kissed him one last time and slipped her backpack into a plastic security bin. Sadly, she waved to them all as she stepped through the metal detector.

"*Haakoh kaama*," she cried, bidding them farewell.

As she approached the boarding steps she shifted the heavy pack from one shoulder to the other. A fatherly looking, gray-haired gentleman smiled at her and said, "Can I take your pack for you?"

She slid it down off her shoulder and put both arms around it. She felt the moccasins where on that first day she had felt her camera.

"No, thanks," she said, "I've got it."

GLOSSARY

The Pueblo language, Keres, is an oral and not a written language. The following words are spelled phonetically to approximate their pronunciation in English.

Dawaa'e: thank you
Dzah: no
Dz'i qwii gudeii kwih: what's up?
Ha'a: yes
Haakoh kaama: goodbye; can I leave?
Heem'e: enough
Henateetz: cloud
Hin'a: okay
Hueh: thank you (male)
Kiva: ceremonial building
Kotra auksi: mean
Kuwe tzi: hello; how are you?
Ma'a: stop
Natchra: thank you (female)
Pakowtza: cookies
Saucha: child
Tiya: dog
Tzanawani: grump, grouchy
Tuu'ma: just kidding
Tzitzi: water
Whooshka: robin

Enjoy an excerpt from

SECOND SUMMER

next in
the Cloud Dancer series

"Erin, if you go back to New Mexico next week, don't expect me to welcome you home with open arms." His gaze held fast to hers. "Honestly, I don't see how you can expect me to welcome you back at all."

"For heaven's sake, Craig, we've gone over this again and again. The plans were made last summer when my cousin, Neyse, was here. It's our family vacation. I'm certainly not staying home while Mom and Dad and my brother go to New Mexico!"

He shook his head in frustration. "Fine. Go."

She didn't know whether to hit him or hug him: frustration or reassurance. Which approach would make things better?

"I'll only be there for two months. I'll be home at the end of August and then we'll go down to Santa Barbara together. We'll be at UCSB for the next two years. Nothing is going to change that."

"Don't count on it," he said stubbornly.

"I *am* counting on it."

"Well, don't. In fact, I don't know why we should keep seeing each other at all. Give your Indian friend a call and ask him to take you to your graduation party. You'd probably have more fun with him anyway."

He stood suddenly, seeming much taller than six feet as he towered over her. "I'm outta here." He stalked across the park to the bike racks.

"Craig," she called, but he kept going. "Craig!"

He never slowed or turned. She watched the back of his head, his brown hair glistening with golden highlights as he unlocked the chain on his bike, wrapped it around the seat post, and rode out of sight.

At first she thought he might come back. When he didn't, she began to feel sick to her stomach. It was always that way when they argued: she'd try to reason; he'd explode and walk away. She'd wait. Usually he'd call later that day or perhaps the next, apologize, and they'd talk it out. She knew that this issue was different. He had never been able to accept the idea that she was going back to the pueblo.

Two years before, when Erin had turned eighteen, she'd spent the summer in New Mexico with her father's sister's family. Her Aunt Julia had married a full-blooded Pueblo Indian, Antonio, and lived there with his people. They had a daughter, Neyse, who was just her age, and a son, Carlos, three years younger.

Her first day there, Erin had met Paolo, her uncle's oldest nephew, who seemed to take an instant dislike to her. Much later, he'd confessed that he had actually been attracted to her those first moments, and was only guarding himself. As the summer progressed, the attraction grew stronger. Erin wasn't sure whether it was love, but she knew it was special. The day she left, Paolo told her, "I'll be here if you come back. I'll be waiting for you."

But despite his promise, Erin had never heard from him again. She'd written letter after letter and never received a single one in return. She'd waited and hoped and made up excuses for him, and never understood. She and Neyse emailed regularly, however Neyse could not explain his actions either. Gradually, she'd pushed him out of her thoughts, dated guys she met at school, and tried to forget.

Neyse had come to California the summer before, and spent two months telling tales of growing up on the pueblo. Erin's parents, and her brother, Scott, adored Neyse, and were so enchanted by the stories, that they promised to visit the pueblo on vacation the following year. That year had come, and the family was ready to leave.

The problem was Craig.

Erin met Craig at a friend's party the past September, and they'd been dating ever since. He'd just finished his sophomore year at the University of Santa Barbara and was home for the summer. At Christmas, when packages arrived from Neyse, Erin told Craig about her summer on the pueblo, and her plan to return. At the time, he hadn't said much about it. As the year had worn on, he had become increasingly upset about her leaving. In the past two days it had blown up into a huge issue, which Erin couldn't seem to diffuse.

Erin had just completed her AA at junior college and was transferring to USCB in the fall. Her graduation was looming, and the trip would commence three days later. She'd tried to reassure Craig that Paolo no longer meant anything to her. It seemed that her protestations had been unconvincing. Although Erin wasn't sure what she felt for Paolo, deep inside she knew she felt something. And no matter what she said, Craig knew it, too.

Her lithe body felt unusually heavy as she stood, brushing her long, honey-colored hair back from her face. *No point in sitting here all afternoon. He certainly isn't going to change his mind this soon.*

Three days later she still hadn't heard from him.

"Why don't you call him?" her mother asked, sitting next to her on the bed after they'd finished the dinner dishes. "All he wants is reassurance."

She felt like screaming. "I've told him a thousand times. What else is there to say?" she asked, her voice carefully controlled.

"Erin, don't let your own stubbornness ruin your graduation. You know he cares for you and wants to be with you, or he wouldn't react so strongly. He's hurting."

"This isn't easy for me either," she said. She was quiet for a while and then she asked, "Do you think I'm wrong to go back there?"

"Wrong in what way?"

"I don't know. Foolish? Unfair to Craig? Asking for trouble?"

"You won't know until you get down there if you've asked for trouble. As far as Craig is concerned, you don't owe him anything except honesty, and you've given him that. I'm just saying that if you'd call him, you might make it easier for him to make amends. I suspect he still wants to take you to the party, but he's put himself in a corner."

Erin started at the sound of the doorbell, and sighed when she heard her father talking to her best friend, Taylor. She sat up, crossing her legs, just as a soft knock came on her bedroom door.

"Come on in, Tay."

Erin's mother stood and kissed her on the cheek.

"Hi, Mrs. Fraser."

"Hello, Taylor. How are you?"

"Fine. Erin…?"

Erin shrugged.

The door closed behind her mother as she left.

"He hasn't called?" Taylor asked.

Erin shook her head.

"Jerk," her friend mumbled, settling herself on the bed against the headboard. "What are you going to do?"

Erin sighed heavily. "I don't know what to do." She looked at her friend who, like herself, was blonde. Unlike Erin, Taylor's hair was a fluffy mass of ringlets. For years Taylor had tried every possible way to straighten it, and only recently had cut it short and let it curl naturally. It was bleached almost white from the sun, as Taylor played tennis nearly every day. She was bigger than Erin in almost every way: tall, full breasted, muscular, both her smile and teeth, wide. While Erin excelled in swimming, Taylor was a second seeded tennis player, and had won an athletic scholarship to Arizona State. Like Craig, Taylor was home for the summer after her second year.

Taylor frowned, her blue eyes concerned. "You don't think he *won't* call, do you? We're talking about your graduation here."

"I don't know what to think. My mom says I should call him."

"Are you going to?"

Erin pulled her knees up to her chest, wrapped her arms around her legs, and rested her chin on her knees. "I don't know. I can't stand to think of graduation without him being there. On the other hand, if we're just going to keep arguing about my trip, what's the point?"

"And you're determined to go?"

"Can you think of a reason I shouldn't? The whole family planned this trip last year with Neyse. She would never understand if I changed my mind now. Craig has known about it for months. I don't know why he is making such a federal case out of it."

Taylor smirked. "I can tell you in one word."

"Everybody acts as if I'm going down there just to see *him*."

"Whom?" Taylor questioned, with mock innocence.

"You know *whom*."

"Say it, Erin. Say his name. He does have a name, right?"

"Cut it out."

"Say it. If he doesn't mean anything to you, you should be able to at least say his name."

"Paolo," Erin said flatly. "Satisfied?"

Her friend shrugged. "I think he deserves that much."

"I haven't heard from him since I left there two years

ago. We only knew each other for two months. For all I know, he's married."

"If he were, I think that Neyse might have mentioned it. He's her cousin too, right?"

"Why should she say anything to me about it?"

"Oh, come on, Erin. It's only obvious that you still think about him. Why don't you admit it?"

"I think about a lot of things."

"Just admit it once and I'll drop it. You've never forgotten him…never gotten over him. You wonder what happened. We both know it, and Craig must know it, too. Be honest with yourself, if not with me."

Erin looked away, out the window toward the dense, dark green hills. *What's the point?* She felt the old ache she'd hoped was gone. Maybe it was a mistake to go back…to open it up again. He obviously didn't care or he would have written. Once, anyway. An explanation of some kind. A reason. *Some*thing. But she hadn't heard a word from him since she'd left, not one word. She thought she'd been special to him. That's what he'd said.

She looked back at Taylor. "Neyse says he doesn't even have a girlfriend. And yet, he refuses to talk about me."

Taylor nodded knowingly. "That's it. Proof. An absolute sign of love."

"Oh, Tay, don't be dramatic."

"Not dramatic…romantic."

"Romantic means long letters, flowers, candy, and every once in a while, a late night phone call. Not hearing from someone for two years is *not* mentioned under the definition of romantic."

"He's waiting for you."

"For me to what? I wrote him so many letters!"

"Maybe they were lost in the mail."

"A dozen letters? Get real."

"A dozen. Now that's romantic," Taylor mused. "Tell me again why you don't email him?"

"He doesn't have a computer. He doesn't want a computer. He hates computers."

"And cell phones."

"Yes, most especially cell phones."

"Well, maybe you had the wrong address."

"That's enough, Taylor. You said you'd drop it. I said his name. Paolo. Paolo, Paolo, Paolo. I admit it. I've never forgotten. But he *has,* and that's what counts."

And don't miss

SUMMER GHOST
third in
the Cloud Dancer series

Erin Fraser and her best friend, Taylor, were on their way to New Mexico, to the Pueblo where Erin had fallen in love with Paolo, a handsome, mysterious, southwest Indian. Erin had had no contact with Paolo for more than four years and wanted desperately to find out what, if anything, might still be between them. She'd accepted a teaching position at a high school near the Pueblo and planned to live there for the next year.

"Have you ever been sorry you didn't transfer to The University of New Mexico years ago?" Taylor asked.

"No. I was happy at UC Santa Barbara," Erin said. "I think I got a really good foundation for teaching. And, Paolo was up at Colorado State anyway. There's no telling if things would have worked out better if I'd transferred to UNM."

Her friend nodded.

"What about you, Taylor?" Erin asked. "Any regrets about college?"

"Only that I didn't study harder. It was all about the tennis back then. I wish I'd paid attention to the academics."

"But the tennis paid the bills," Erin said, reminding her of her athletic scholarship.

"Still does. Oh, and, I wish I'd gone to my graduation," she added. "At the time I thought it was corny and I just wanted to get done and get out. Now I wish I'd gone."

Erin shook her head. "And I almost wish I hadn't."

"What?"

"I made a fool out of myself. I never told you, or anyone for that matter."

"What the heck happened?"

"We were marching in and I tripped somehow. I not only fell over, I knocked over the guy in front of me. My parents didn't see it, thank heaven. I was out of their view. But it was mortifying. I was so shaken up that I could hardly make it across the stage to get my diploma."

"And you never told me," Taylor said sounding hurt.

"At first I felt so stupid and then I tried to forget it."

"Well, if nobody saw it, why does it still bother you?"

"Half the audience saw it—just not the half where my family was sitting."

"Well, if nobody gave you any grief, then I think you can probably forget it."

Erin nodded and after a bit she said, "I think you're right. What a dumb thing to hold on to." She shook her head. "I'll bet no one else noticed, much less remembers."

They drove to Needles in ten hours, spent the night in a motel and completed the trip the next day. Since Taylor had gone to school at Arizona State, she was familiar with the desert landscape. Erin hadn't seen the stark, barren terrain in four years and was surprised at how deeply the beauty of it affected her.

As they neared the pueblo, Erin's stomach began to churn. "I think I'm getting cold feet, Tay," she said. "I'm not so sure this is such a great idea."

"Don't you dare back out now," Taylor threatened. "This is something you have to do...whatever the outcome."

Erin nodded reluctantly. "I know you're right. You are. But right at this moment," she paused, "I'm scared."

"There's only one way to do this and that's in person.

You have to go."

Erin swallowed hard. "I know I do," she said, her voice gaining resolve. "I need to find out once and for all."

They stopped and called Neyse on Erin's cell. "We're in Albuquerque. Do you need anything?"

"No. And for heaven's sake don't stop to eat. Mom has enough food for an army."

"Surprise, surprise. Okay, we'll be there soon."

"Come straight to Mom's. We'll take you home after you've visited with the folks a while."

"Right. See you."

When they turned off the highway to the pueblo, Taylor pointed at the big sign there listing all the things prohibited in the village. "The famous sign."

Erin nodded. "That's it." She laughed. "It's been six years since I saw it that first trip. It's hard to believe."

The road to her aunt's house was as familiar as if she'd driven it the day before. She drove slowly across the canal so Taylor could see the beautiful, clear water. They were soon pulling into the yard in front of Julia and Tony's home. She hadn't even turned off the engine when the front door opened and her family poured out.

"Welcome," Julia said, hugging them both. "I'm so glad you're here."

Neyse, Carlos, and Tony also greeted them. Then Juan came out on the porch.

"*Kuwe tzi*," he said.

Erin nodded and replied, "*Kuwe tzi*," telling him hello in his native tongue.

Taylor had met everyone except Juan when they'd visited California years before. Erin pulled Juan over to her and said, "Taylor, this is my cousin-in-law, Juan."

They all laughed.

"Enough already," Carlos said. "Let's eat."

Julia showed Taylor to the bathroom so she could wash up but Erin headed straight for the table.

"Oh, Auntie," she exclaimed, "prune pie!"

She grabbed a piece of her favorite pueblo treat and took a huge bite. She moaned. "It tastes just as good as I remember."

Julia beamed.

As they all ate, the girls answered questions about their trip and their plans for the next few days.

"I want Taylor to see *everything* before she has to leave," Erin said. "The canal, the river, the garden, Tent Rocks, the lake, the rec center…everything!"

"There aren't enough days for all that," Juan said. "You'll have to do some careful planning."

"You have another more important task, I think," her uncle said. "You need to find a place to live. Are you on the housing committee, Taylor?"

"Absolutely," Taylor said.

"I don't know how we'll manage it all," Neyse said, "but we will. It'll work out."

"If we don't get it all done in time," Erin said, "she'll just have to call in sick."

"Right," Taylor said. "That'll fly."

"Not," Erin said with a grin.

They stayed and talked for a while after dinner until Juan said, "I think we should get home so we can unload the car while we still have some light." Erin hugged her aunt and uncle and promised to visit again before Taylor left. Then, Juan led the way to his and Neyse's house.

It was a newer pueblo home, not unlike Julia and Tony's, with three small bedrooms and one bathroom, fairly standard on the pueblo.

"I fixed one room for you two to share, if that's okay," Neyse told them.

"If only Erin didn't snore," Taylor teased.

Erin hit her. "Do not! Why don't you take a shower, Taylor, and I'll do some unpacking."

"Sounds good to me," Taylor agreed.

Juan disappeared and gave Erin and Neyse their first moments alone since the girls had arrived.

"This is so great," Neyse said. "You're here to stay!"

"For a year anyway. After that we'll just have to see."

"What made you decide to come back?"

"Lots of things. I've never gotten the pueblo out of my blood, I guess."

"Is that all?" Neyse asked, eyeing her keenly.

"I suppose not," she said quietly. After a time she asked, "Does he still live here?"

Neyse nodded. "With Mr. Montoya, actually. The old man is pretty feeble and Paolo helps him out."

"So, he's not married then."

"No."

"Has he ever asked about me?"

Neyse hesitated. "Just once."

"When was that?"

"Last year. When you graduated. He wanted to know when it was. He told me he was just curious. But he never mentioned it again and he's never asked me another thing. Strange, don't you think?"

Erin was quiet, a soft smile on her face.

Neyse looked at her, surprised. "You've never gotten over him," she said in amazement. "Oh, Erin, I'm so sorry."

Erin chilled at the sound of her voice. "Why?" she asked, alarmed.

Neyse shook her head sadly. "He's been dating Natoma again."

Natoma, the beautiful girl he'd dated years before. Of course he'd be dating. She hadn't even considered it. Erin felt the solid earth slipping out from under her feet and had to concentrate on keeping her balance. All this time…all this way…only to hear he was with Natoma.

Years of controlling her tears served her well as she shrugged her shoulders. "I guess it wasn't meant to be."

But that night, long after the house was silent, tears of bitter regret streaked down Erin's cheeks. *He's been dating Natoma again.*

Why? Why did I wait so long?

Linda McGinnis
is the author of the *Cloud Dancer* series,

She also wrote
Dreams of Home, chronicling the struggles
of two Japanese families in Hawaii
during World War II.

Coming in the spring of 2014:
Till I Kissed You.
This is the first of the Sweet Refrain Series,
a vintage new adult novel
about a young woman who attends an
all-women's college in the 1950s.

When not writing,
Linda enjoys traveling, photography
and quilting.

Visit her soon on the web at
LindaMcGinnis.com

Made in the USA
Las Vegas, NV
19 December 2022

63457753R00148